About the Author

Dave Emanuel began his career in automotive journalism in 1970 when his first article was published in "Car Craft" magazine. Since then, his byline has appeared on over 1,500 feature articles in magazines such as "Road & Track," "Hot Rod," "Chevy High Performance," "Super Chevy," "Stock Car Racing," "Circle Track," "Drag Racing Monthly," "Bracket Racing USA," "Super Ford," "Popular Hot Rodding," "Motor Trend," "Corvette Fever," "Muscle Car Review," "Home Mechanix," "Popular Science," "Automotive Rebuilder," "4-Wheel & Off-Road," "Open Road" and "Automobile Quarterly." He has also written six books and has appeared in a number of instructional videos, and on the nationally televised program "Road Test Magazine" on The Nashville Network (TNN). A thorough knowledge of a subject is required before an author can write authoritatively about it and Dave gained his knowledge about automobiles the old fashioned way — he broke them. Even before he acquired a driver's license, Dave experienced the joy of popping the clutch and the agony of replacing the transmission. And it wasn't long after he began driving legally that he made his first visits to a drag strip, initially as a spectator, then as a competitor. After several years of racing experience, the marriage of journalism and automobiles came about when Dave set a track record and decided to chronicle his efforts After dabbling in journalism on a part-time basis for several years, Dave decided to abandon his job as a computer systems analyst and began writing full time.

In a relatively short time, he established himself as one of the country's leading automotive journalists. Although he has done road tests, product reviews, personality profiles and features, he is best known for his technical articles and books. He brings a unique perspective to his work through a combination of personal relationships with a number of the most successful race engine builders in the nation, and extensive hands-on experience.

This book is an independent publication, and the authors and/or the publisher thereof are not in anyway associated with, and are not authorized to act on behalf of, the carburetor manufacturer and distributor, Holley Performance Products.

Further, the above-mentioned Holley Performance Products did not sponsor and do not endorse the contents of this book and are in no way affiliated or associated with the publication thereof. Since certain carburetor modification processes discussed herein may void the Holley Limited Warranty, such Holley Performance Products can in no way be held responsible for such modifications performed by the customer/reader.

Holley® and Dominator® are trademarks of Holley Performance Products. The publisher has been granted the limited rights to copy from Holley publications and use certain pictures, illustrations and other proprietary material of Holley Performance Products contained herein; otherwise, all rights in such materials are reserved by the copyright owner, Holley Performance Products.

The text, photographs, drawings and other artwork (hereafter referred to as information) contained in publication is sold without any warranty as to useability or performance. In all cases, originals manufacturer's recommendations, procedures and instructions supersede and take precedence over descriptions herein. Specific component design and mechanical procedures — and the qualifications of individual readers are beyond the control of the publisher, therefore the publisher disclaims all liability, either expressed or implied, for use of the information in this publication. All risk for its use is entirely assumed by the purchaser/user. In no event will S-A Design be liable for any indirect, special or consequential damages, including but not limited to personal injury or any other damages, arising out of the use or misuse of any information in this publication.

The publisher reserves the right to revise this publication or change its content from time to time without obligation to notify any persons of such revisions or changes.

Super Tuning and Modifying
HOLLEY® CARBURETORS

EDITED BY
JEAN BOWERMAN

PRODUCED BY
TAMARA BAECHTEL

TJ 787 .E43 1999

Emanuel, Dave, 1946-

Super tuning and modifying Holley carburetors

Copyright © 1999, 1988, 1983 by Cartech, 11481 Kost Dam Road, North Branch, MN 55056. All rights reserved. All text and photographs in this publication are copyright property of CarTech. It is unlawful to reproduce — or copy in any way — resell or redistribute this information without the express written permission of the publisher. Printed in USA.

ISBN 1-884089-28-3
Part No. SA08

DELTA COLLEGE LIBRARY

CARTECH • 11481 KOST DAM ROAD• NORTH BRANCH, MN 55056

CONTENTS

1. INTRODUCTION 4
2. BASIC MODELS AND PART NUMBERS 8
   - THE BASIC MODELS 8
   - EVOLUTIONARY CHANGES 10
   - UNDERSTANDING HOLLEY PART NUMBERS ... 11
3. SELECTING THE RIGHT HOLLEY CARB 14
   - DETERMINING FLOW REQUIREMENTS 15
   - VACUUM OR MECHANICAL SECONDARIES ... 18
   - CARB/MANIFOLD/CAMS INTERRELATIONSHIPS ... 18
   - OTHER CONSIDERATIONS 19
4. BASIC FUNCTIONING 20
   - PRIMARY METERING SYSTEMS 20
   - POWER ENRICHMENT 25
   - ACCELERATOR PUMP 27
   - IDLE SYSTEM 30
   - REVERSE IDLE SYSTEM 33
   - CHOKES AND COLD STARTING 35
     - SECONDARY METERING SYSTEMS 37
     - SECONDARY IDLE CIRCUIT 37
     - SECONDARY MAIN METERING CIRCUIT ... 37
     - POWER ENRICHMENT 37
     - SECONDARY ACTUATION 37
     - MECHANICAL SECONDARIES 38
   - FUEL INLET CONTROL 38
     - FLOAT BOWLS 38
     - SETTING OF FUEL LEVEL 39
5. BASIC MODIFICATION CONCEPTS 40
   - PRIMARY IDLE SYSTEM — STANDARD 40
   - PRIMARY IDLE SYSTEM — REVERSE 42
   - MAIN METERING SYSTEM 43
   - AIR BLEED MODIFICATIONS 46
   - POWER ENRICHMENT CIRCUIT 47
   - ACCELERATOR PUMP CIRCUIT 51
   - VACUUM-ACTUATED SECONDARIES 53
   - MECHANICALLY ACTUATED SECONDARIES ... 54
   - FUEL INLET SYSTEM 55
   - TUNING FOR HIGH ALTITUDE 56
   - GENERAL TUNING TIPS 56
     - FILTERS 56
     - CARB HEAT 57
     - COLD 57
     - LEAKAGE 57
6. MODS FOR STREET PERFORMANCE 58
   - MILLING OF FLAT SURFACES 58
   - FUEL BOWL VENT BAFFLING 59
   - OPTIONAL FUEL BOWLS 60
   - SECONDARY ACTUATION 61
   - ACCELERATOR PUMP 62
   - POWER VALVES 62
   - STREET JETTING 63
   - DOUBLE-PUMPERS FOR THE STREET 64
7. MODS FOR TURBOCHARGING 66
   - DRAW-THROUGH MODIFICATIONS 67
   - BLOW-THROUGH MODIFICATIONS 68
8. MODS FOR COMPETITION 72
   - CHOKE SHROUD (HOUSING) REMOVAL ... 73
   - MACHINING OF FLAT SURFACES 74
   - ACCELERATOR PUMP MODS 75
   - THROTTLE BODY SUBSTITUTION 76
   - PRIMARY IDLE CIRCUIT MODS 78
   - SECONDARY IDLE CIRCUIT MODS 79
   - MISCELLANEOUS IDLE CIRCUIT MODS ... 81
   - POWER ENRICHMENT CIRCUIT MODS 83
   - FLOAT BOWLS 84
   - MODIFIED AIR BLEEDS 84
   - THINNING OF THROTTLE PLATES & SHAFTS ... 85
   - INTERMEDIATE CIRCUITS 85
   - MODIFICATIONS FOR ALCOHOL 85
9. THE MODEL 4500 88
   - AIR HORN EXTENSION 91
   - MACHINING FLAT SURFACES 92
   - ACCELERATOR PUMP 92
   - FLOAT BOWLS 92
   - MAIN METERING SYSTEM 92
   - INTERMEDIATE SYSTEMS 94
   - AIR BLEEDS 95
   - POWER ENRICHMENT 95
   - LINKAGE 95
   - BOOSTER VENTURI 96
   - IDLE FEED RESTRICTIONS 97
10. ECONOMY & SPECIAL APPLICATION TIPS ... 98
    - ECONOMY 98
    - OFF-ROAD 102
    - MARINE 104
11. THE MODEL 4360 106
    - BASIC FUNCTIONING 107
    - MODIFICATIONS 110
    - POWER VALVE 110
    - ACCELERATOR PUMP 110
    - FLOAT LEVEL 111
12. REBUILDING & TUNING MODELS 4150/4160 ... 112

APPENDIX 120
   - EXPLODED VIEW, MODEL 4500 120
   - EXPLODED VIEW, MODEL 4360 121
   - EXPLODED VIEW, MODEL 4150/4160 122
   - EXPLODED VIEW, MODEL 2300 124
   - NUMERICAL LISTING 126
   - HOLLEY REFERENCE CHARTS 144

OTHER S-A DESIGN BOOKS 145

Super Tuning and Modifying HOLLEY CARBURETORS

Introduction

The Model 4500 Dominator is the largest-capacity automotive carburetor ever produced. In both airflow capacity and sophistication, it's light years ahead of the single-barrel carburetors that Holley produced when the company was first founded.

For over 30 years the name "Holley" has been synonymous with top caliber, high-performance automotive carburetion. However, the Holley performance image was a bit blurry until the late 1950s when it was brought into sharp focus by the Model 4150 four-barrel carburetor. In 1957, Ford Motor Company became the first automaker to take advantage of the new carburetor when a 400cfm version appeared on the upgraded, high-performance 312 engine. By 1964, General Motors and Chrysler had seen the error of their collective ways and began using Holley four-barrels as original equipment carburetors on selected high-performance engines.

But Holley was not always at the apex of performance carburetion. In fact, the company's genesis was in no way related to carburetors of any type. In 1896 George M. Holley, a man who loved to tinker with things, married a small gasoline engine to the frame of a bicycle, thereby creating the first American motorcycle. A year later, at the tender age of 19, Holley designed and constructed a three-wheeled, chain-driven car powered

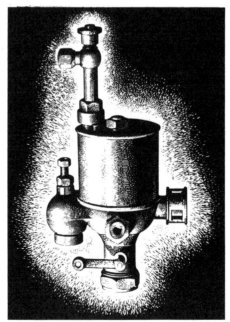

The Holley brothers began to specialize in carburetors with the introduction of the "Iron Pot," designed for the 1904 "curved-dash" Oldsmobile.

by a single-cylinder, gasoline-burning engine. Seated behind the tiller of the three-wheeler, Holley could blaze through his native Pennsylvania countryside at the incredible speed of 30 miles per hour. In addition to making all the patterns and castings for parts used on the car, Holley created a carburetor design that was to serve as the basis for future development.

George Holley developed a tremendous fascination with the mechanics of speed and soon began racing motorcycles in national competition. But George had other aspirations and in 1899, he and his brother Earl formed the Holley Brothers Company for the purpose of manufacturing motorcycle engines and, ultimately, motorcycles.

The first was completed in 1901. George remained an active competitor. In 1902, he won the first American Motorcycle endurance contest and set a number of world speed records at the Pan American Exposition in Buffalo, New York.

In 1903, the Holley brothers, having recently added two wheels to their designs, introduced the Holley Motorette, a sprightly little runabout, that boasted a 5-1/2 horsepower engine, planetary transmission, and tilt steering wheel.

George Holley prepares for a ride in a replica of the chain-driven, three-wheeled car that he built in 1897 when he was 19 years old.

The Holley Motorette, built between 1902 and 1906, sported a 5.5 horsepower motor, planetary transmission, and tilt steering wheel.

Over 600 Motorettes were produced between 1903 and 1905, but few remain today. (However, one is always easily found on display at Holley's corporate headquarters in Bowling Green, Kentucky.)

The year 1904 was a milestone in the history of the Holley name as the company made its first move toward specialization with the introduction of the "Iron Pot" carburetor for the now-famous curved-dash Oldsmobile. Henry Ford also used the "Iron Pot" on his Model T, and Holley carburetors soon became original equipment on Buick, Winton, and Pierce-Arrow automobiles, as well.

The manufacture of Motorettes and "Iron Pots" were concurrent efforts and the Holley brothers rapidly learned that greater overall success could be achieved if they concentrated on a single endeavor. According to an old story of questionable veracity, Henry Ford, apprehensive of the potential competition represented by George and Earl Holley, helped the brothers with their decision to specialize in carburetors. Ford supposedly made the Holleys a simple proposition: "You don't build cars and I won't build carburetors." Whether the story is true or fabricated is a matter of speculation, but, by 1906, Motorette production had ceased and the carburetor business had begun to boom. As years passed, Holley carburetors became original equipment on an ever-increasing variety of cars, including the Model A Ford, for which some 16 million carburetors were built. Since 1904, Holley has built over 100 million carburetors. (By the way, Henry Ford never did build carburetors.)

Holley continued to supply Ford Motor Company and other automakers with one- and two-barrel carburetors through the 1940s, but with the introduction of the 1953 Lincoln, something new was added — the Model 2140, a 370cfm four-barrel. Highly unusual in appearance, this carburetor, like the Model 1901 two-barrel from which it was

An improvement over the "Iron Pot," the Holley Model NH carburetor was standard equipment on the Model T Ford for several years.

derived, had the float chamber built into the main body casting.

In 1955, the Model 4000 four-barrel (also rated at 370cfm) made its debut atop Ford's entry in the Detroit horsepower sweepstakes—the 292 cubic-inch "Y" block. During the following year, Ford used the strange-looking Model 4000 (most memorable because of its "trap-door" choke plates) in single and dual four-barrel installations, but by 1957 the new higher capacity Model 4150 (400 as opposed to 370cfm) had replaced the older design in all applications except the 265-horsepower, dual four-barrel-equipped 312.

With its unique modular design, generous airflow capability and superb fuel metering characteristics, the 4150 was initially surrounded by an aura of mystery. But when a piece of equipment offers exceptional performance potential, whatever aura surrounds it soon evaporates like gasoline on hot pavement. It took just a few years for the Holley Model 4150 to become the world's most popular performance carburetor.

In a sense, Holley's modular four-barrel is the carburetor equivalent of the small-block Chevy engine. Both were designed to satisfy demands that have changed dramatically over the years, and both have proven to be sufficiently versatile that 40-odd years after their inception, they're still a prickly thorn in the side of their competition. Model 4150 Holley

Super Tuning and Modifying Holley Carburetors

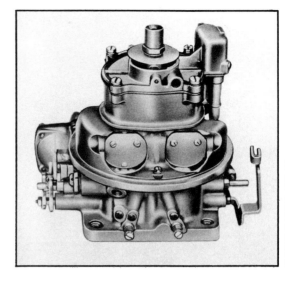

With the advent in 1953 of the Model 2140, Holley moved into the realm of four-barrel carburetion.

This carburetor, original-equipment for the Model "A" Ford, probably holds the record, at over 16 million, for the greatest number produced.

It's hard to believe, but this is the four-barrel design that immediately preceded the Model 4150. It is a Model 4000 and was used on Ford engines through 1957.

Originally introduced in 1957, the Holley Model 4150 defines the term "performance carburetor." No other carburetor has been used in such a variety of performance and racing applications. At left is a "double-pumper"; on the right is a version with vacuum secondaries.

four-barrels, essentially the same carburetors that flowed new life into the Ford "Y" block way back in 1957, are still found on some of the most advanced racing engines in the world.

But at the original equipment level, a dramatically different scenario unfolded. As government emissions regulations began their inexorable encroachment into Detroit engineering laboratories, the automakers' ability to market high-performance engine options diminished. Holley continued to supply original equipment carburetors through the 1980s, but the days of factory-installed 780- and 850-cfm four barrels had come to an end.

To the nation's new ecological awareness, Holley responded with a variety of "emissions design" replacement carburetors. The venerable 4150/4160 design again proved its timelessness as Holley engineers formulated emissions calibrations for updated models, with airflow capacities ranging from 600 to 780cfm. In 1971 Model 4165/4175 Spread-Bore carburetors were introduced as direct bolt-on replacements for the anemic Rochester Quadrajets. In typical Holley fashion, The Spread-Bores provided both improved performance and increased gas mileage while retaining original exhaust emissions levels.

Emissions requirements for the millions of single- and two-barrel carburetors Holley produced each year required a large amount of engineering effort during the late 1960s and early 1970s, but this did not preclude additional development in a highly non-emissions area — race-car carburetion. As camshaft and cylinder head technology advanced, large-displacement race engines began developing shortness of breath when fitted with an 850cfm carburetor. Ford Motor Company's involvement in NASCAR and Trans/Am racing was far from casual during the latter 1960s so the company contracted with Holley to design and produce a carburetor with greater airflow capacity. The result was the 1969 release of the first Model 4500 "Dominator" four-barrel, rated at 1150cfm. Originally dubbed the "mystery" carb, initial use was restricted to Ford 429 Nascar and 302 Trans/Am engines, but Dominators rated at 1050 and 1150cfm are now used in a variety of single and dual four-barrel racing applications.

The next new carburetor design produced by Holley was the Model 4360 four-barrel. a low cost Quadra-Jet replacement designed for V6 and low horsepower V8 engines. Introduced in 1976, the Model 4360 was a low-cost Quadra-Jet replacement rated at 450cfm and calibrated for exhaust emissions/gas-mileage-critical applications. For a time, the future looked promising for the 4360, but when contrived oil shortages ended, demand

The 396-454 big block was the first Chevrolet engine family to receive Holley carburetion on all high-performance models. Airflow ratings ranged from 585 to 830cfm.

Originally designed to give Ford an advantage on the super speedways, the Model 4500 is today the ultimate carburetor for professional drag racing.

A baby brother to the 4150, the Model 2300 two-barrel also has been used in a diverse array of applications. The "race-only" design flows 650cfm. The standard version offers less than half that capacity.

The Holley Model 4165/4175 Spread-Bore was created as a high-performance replacement for the lackluster Rochester Quadra-Jet. It provided all the advantages of the Holley modular design while retaining the spread-bore mounting flange used on many stock manifolds.

The Model 4010 and 4011 carburetors were introduced as "show" carburetors in 1990. Functionally almost identical to the 4150/4160 models, the 4010/411 series employed completely different construction. (Only two major castings: the main body and air horn, both aluminum.) These carburetors were discontinued in 1998.

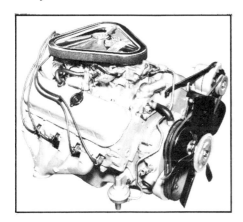

One of the last factory engine options to feature multiple carburetors was the 427 Corvette powerplant of 1969. It was rated at 435 horsepower and featured three Holley Model 2300 two-barrels.

Introduced in 1976, the Model 4360 was an emissions-era carburetor designed as a low-cost replacement for the Rochester Quadra-Jet. It was discontinued in the early 1990s.

A splash of underhood chrome easily can be added to any engine with one of Holley's 4150/4160 show carburetors. Individual chrome parts are also available separately.

surged for carburetors with higher air flow capacities. Concurrently, stricter exhaust emissions standards prompted a move toward electronic fuel injection.

Holley responded with a variety of traditional modular carburetors, new non-modular design, Model 2010, 4010 and 4011 carburetors, replacement high capacity throttle bodies for electronic TBI systems and Pro-Jection, a stand-alone throttle-body-style fuel injection system.

The world has change dramatically since 1957, but through feasts and famines, good times and bad, Holley Performance Products, as the company is now known, has risen to the challenges. Undoubtedly, it will continue to do so well into the 21st century.

Super Tuning and Modifying Holley Carburetors **7**

Super Tuning and Modifying HOLLEY CARBURETORS
Basic Models and Part Numbers

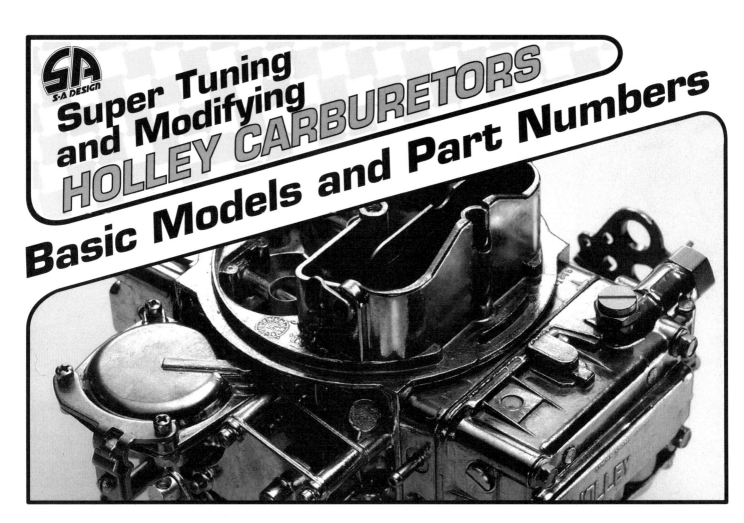

The Holley Model 4150 and 4160 four-barrels and the Model 2300 two-barrel are closely related members of the same carburetor family. The Model 2300 two-barrel is nothing more than the primary two-barrel section of a Model 4160 four-barrel. The 4150 and 4160 are virtually identical to each other with the exception of the hardware used to meter fuel in the secondary idle and main circuits. Rather than employing a block with removable jets (as used on the primary side), 4160 carburetors are fitted on the secondary side with a metering plate that contains non-replaceable fuel metering restrictions. The Model 4150 has metering blocks with replaceable jets in both the primary and the secondary circuits.

Model 4165 and 4175 Spread-Bore carburetors, although very similar in appearance to the previously mentioned 4150/4160 models, differ considerably in construction. These carburetors are more like cousins than brothers. While the family resemblance is indeed strong, the Spread-Bores share surprisingly few components with their relatives, in spite of the fact that fuel metering circuits are conceptually identical. Power valves, jets, fuel bowl vent accessories, secondary diaphragm springs, floats, needles-and-seats and some accelerator pump hardware are the only items that will successfully interchange between the 4150/4160 and the 4165/4175 series.

The reasons for the uniqueness of 4165/4175 components are both logical and straightforward. These carburetors were designed as direct bolt-on replacements for the Rochester Quadra-Jet and Carter Thermo-Quad, and as such must function in an emissions-sensitive environment. Carburetor components for the 4165/4175 family were, therefore, designed to provide improved driveability while holding exhaust emissions within acceptable limits.

In 1976, Holley introduced a model 4360 four-barrel, a low-cost Quadra-Jet replacement. Designed primarily to meet "Energy Crisis" era demand for fuel economy, the 4360 lived a successful, albeit short, life. As fuel shortages eased and electronic fuel injection came to the fore, the 4360 faded from sight.

But, in a sense, its spirit lived on in the model 2010, 4010 and 4011 carburetors that debuted in 1990. Like the 4360, these carburetors are of the non-modular design and have float bowls incorporated in a large aluminum main body casting. Model 4010 has a standard square mounting flange, produced in 600cfm or 750cfm versions with either vacuum or mechanically actuated secondaries. Model 2010 is a two-barrel equivalent and Model 4011 is a Spread Bore version with either 650cfm or 800cfm air flow capacity and vacuum or mechanical secondary actuation.

In spite of their unique construction, Model 4010/4011 carburetors are conceptually very similar to the 4150/4160 series and use many of the same service parts including jets, power valves, needle and seat assemblies, accelerator pump cams and diaphragms.

Identification of the various Holley four-barrel models is quite simple:

a) Model 4150 contains a thick secondary metering block virtually identical to the one used on the primary side, with the exception that it lacks adjustable idle-mixture screws (although some race carburetors do have secondary idle-mixture screws).

b) Model 4160 uses a thin secondary metering plate that fits entirely within the float bowl and is therefore not externally visible. The jets in this plate are not replaceable.

c) Model 4165 employs mechanical secondary actuation, dual accelerator pumps and a secondary metering block with removable jets (similar to the one used on 4150s.)

d) Model 4175 is available only with vacuum operated secondaries and is fitted with the same drilled restriction metering plate that is used on 4160s.

e) Model 4010 is available with either vacuum or mechanically actuated secondaries and fits standard square-flange intake manifolds.

f) Model 4011 is a Spread-Bore equivalent of the 4010 and is also available with either vacuum or mechanically actuated secondaries.

While the carburetors classified as Models 4165 and 4175 are all rather similar, the same is not true of 4150s and 4160s. Within the latter series are four-barrel carbs with: dual accelerator pumps and mechanical secondaries; single accelerator pump and vacuum secondaries; single "centersquirter" accelerator pump with mechanical secondaries; adjustable idle screws on primary only; adjustable idle screws on primary and secondary; center inlet fuel bowls; standard idle mixture adjustment; and reverse idle mixture adjustment. The combination of items seems to be endless in this series, which offers various airflow capacities ranging from 390 to 855 cfm.

In comparison, 4165/4175 carbs are offered in either 650 or 800 cfm versions; all have 1-5/32-inch diameter primary venturi, 1-3/8-inch primary throttle bores and 2-inch diameter secondary throttle bores. The 800 cfm model's extra airflow capacity is derived from 1-23/32-inch secondary venturis, compared to the 1-3/8-inch venturis used on the smaller 650cfm carb. Some 4165s use the reverse-idle system and two types of float bowls are also available. These are the only major differences between the Spread-Bores.

Perhaps the ultimate factory performance/racing induction system — the optional dual four-barrel manifold with Holley carburetors — was offered for the 302ci Z-28 Camaro engine. This configuration was never factory installed, but was available through dealers.

The multiplicity of 4150/4160 configurations can cause initial attempts at understanding the series to be somewhat frustrating. However, if one remembers that the only real difference between the 4150 and 4160 is the secondary fuel metering arrangement, and that a 4150 may be converted to a 4160 (rarely done) and vice versa, it can be seen that all other component variations are analogous to new car optional equipment — the basic unit stays the same,

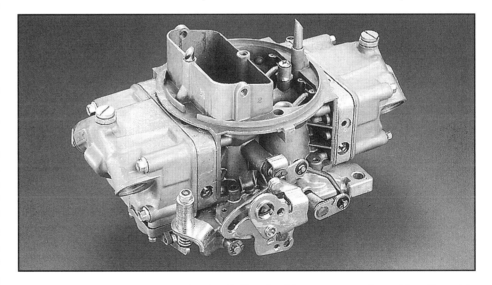

The definitive race carburetor is the Holley double-pumper series with adjustable secondary-idle circuit. Newer models also feature annular discharge booster venturis. However, for street applications, nothing can match a 4150 or 4160 with vacuum-actuated secondaries.

Super Tuning and Modifying Holley Carburetors **9**

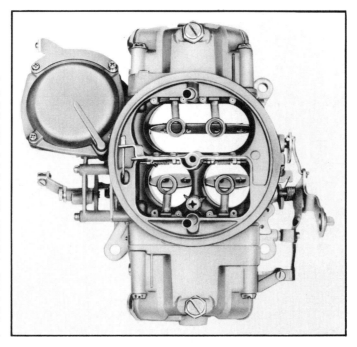

A Model 3160? You bet. There actually was a Model 3160 and it was a three-barrel carburetor. It was introduced for use on the large-displacement engines of the '60s and was listed under the part number R-3916-1AAS. The rear barrel measured 1 3/4 x 3 5/8 inches and airflow was 950 cfm.

The Model 4150 has been manufactured in many different configurations with a variety of float bowls. The Le Mans bowl pictured has been dropped, as it offered no advantage over the center inlet, "dual feed" bowl.

Basic functioning of the Model 4150 and the Model 4160 is virtually identical. However, secondary main fuel metering in the 4160 (right) is controlled by a "metering plate" contained entirely inside the secondary float bowl. In the 4150 (left), it is contained in a "metering block" sandwiched between the bowl and the main body.

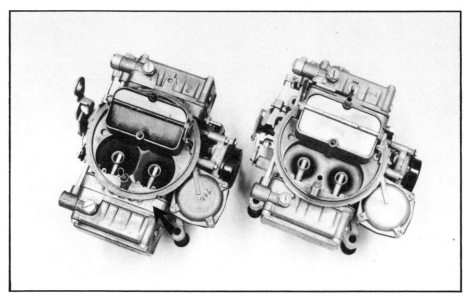

even though options make extensive differences in appearance.

EVOLUTIONARY CHANGES

Since it was designed in the late 1950s, the 4150/60 Holley four-barrel has had to undergo continual engineering development in order to meet changing priorities. As time progressed and emissions requirements became more stringent, special calibrations were developed for street applications. Concurrently, auto racing became highly specialized, creating a need for rather unique carburetion. These two areas are diametrically opposed to one another, necessitating the great variety of part numbers that are now available.

Most early list numbers are vacuum secondary carburetors with standard idle adjustment. Many of these were developed during the "Muscle Car" era as original equipment items and therefore serve as excellent performance carburetors even today. Some, like the 0-3310, have been modified for universal application. The 3310 has proven to be such a popular carburetor that Holley developed a 4160 version (0-3310-2) in an effort to keep the price as low as possible.

The first "double-pumpers" were also developed during the time of Detroit's heavy involvement with performance and racing. Part number 0-4296 was originally offered on certain big-block Corvette engines and included staggered jetting to improve cylinder-to-cylinder fuel distribution. But the double-pumpers that are most popular today evolved somewhat later and were designed as universal carburetors. The 4700 series, ranging from 0-4776 to 0-4781, is comprised of six carbs, with airflow capacities increasing from 600 to 850 cfm in 50 cfm increments.

Designed strictly for drag racing, the "center-squirter" is available in a 660 cfm capacity as part 0-4224. These carburetors use a large 50cc accelerator pump and centrally located pump discharge nozzle that dispenses fuel into all four barrels simultaneously.

The reverse idle carburetors are typically 600 cfm, 4160, vacuum secondary four-barrels calibrated for use on specific late-model "emissions" engines. Most have electric chokes for ease of installation.

In 1978, Holley introduced another variation of the 4150. Two part numbers, 0-8156 and 0-8162, were released specifically for single four-barrel drag race applications. Adjustable secondary idle mixture capability was included to provide increased idle airflow, thereby providing greater compatibility with engines fitted with long-duration, high-overlap camshafts.

For the sake of brevity, future technical discussions will center around the Model 4150/4160 carburetors. It can be assumed that this information will similarly apply to the 4165/4175 models, which are very similar to the 4150/4160 in nearly all basic functions. Specific reference to the Spread-Bore models will be made only when pertinent information specifically does not apply or only applies to these models.

UNDERSTANDING HOLLEY PART NUMBERS

It doesn't take many trips through the Holley catalog to discover that there appears to be some amount of logic to the part numbering system. Unfortunately, the rhyme and reason are somewhat elusive. Sometimes, a designation may be totally numerical; others will be alphabetical and numerical with or without an alpha, alphanumerical or numerical suffix. Confusing, to say the least.

The irregularity among the designations is due to the advent of the much-heralded, solve-the-problems-of-the-world computer. Originally, Holley used alphanumeric part numbers such as R-1850AAS, but the computer found such a bizarre assortment of letters and numbers unpalatable. Hence, the "R" was replaced with an "O," and the AAS, which previously indicated an aftermarket replacement carb, was deleted. The computer swallowed it, so R-1850AAS became simply 0-1850. In most cases, "R" prefixes were converted to zeroes, which signify a high-performance carburetor. In the majority of the remaining cases the number will begin with a "1" prefix. Formerly referred to as a sales number, this is now used pri-

The Holley Model 4165 brought the undisputable advantages of the revolutionary Holley "double-pumper" concept to the Spread-Bore family of carbs. Though a direct replacement for stock low-performance carbs, the 4165 produced neck-snapping performance with factory-type manifolds, precluding the necessity of a manifold swap.

The venerable Holley 3310 dates to 1965, when it was used as an original-equipment carburetor on the 425 horsepower/396ci Chevrolet. Originally rated at 780 cfm, this carburetor has since been revised several times for universal installations. It is currently rated at 750 cfm.

The difference between a double-pumper and a vacuum secondary four-barrel easily may be seen in this photo. Note great similarity between float bowls on the secondary side. The basic casting is identical, but since no accelerator pump circuit is used with vacuum actuation (left), the pump hardware is not installed and the fuel supply passages are not drilled.

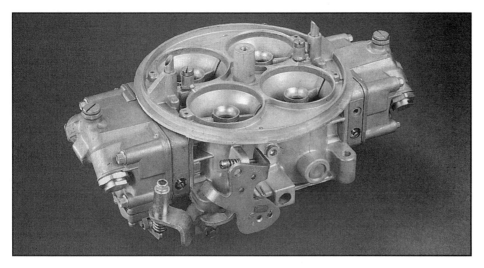

The Holley Model 4500 is also referred to as the "Dominator." Several versions are available, but this carburetor is best suited for racing applications. It will not fit on standard Holley mounting flanges. Race versions are rated at 1050, 1150, and 1250 cfm. A street version rated at 750 cfm is also available.

Model 4165/4175 Spread-Bore carburetors were designed specifically for engines like this Pontiac 455, which were originally factory equipped with a Rochester QuadraJet. The Spread-Bore models are direct replacements for QuadraJets.

The Model 2300 Holley is basically identical to the primary portion of a Model 4150. This carb dominates many forms of two-barrel racing (when non-stock carburetion is allowed) and also can provide outstanding street performance.

marily to signify an OEM service replacement as opposed to a non-OEM high-performance carburetor. (The abbreviation "OEM" stands for Original Equipment Manufacturer. Prior to the advent of electronic fuel injection, "OEM" carburetors were originally supplied to General Motors, Ford or Chrysler for use on a stock engines installed in new vehicles.)

Inconsistencies in the numbering system are not unusual, and arise from the fact that the same carburetor may have been used in a variety of applications. The 0-1850 is a perfect example. It was initially an original equipment carb on 1958-1961 Lincolns with 430 cubic-inch engines, but it also achieved great popularity as a performance replacement carb. (It is still one of Holley's fastest selling numbers, even though the original application is over 40 years old.) If this carb hadn't achieved such wide aftermarket acceptance, it would have existed only as an OEM service replacement for a few years, and then would have been dropped from the catalog (as the 430 Lincoln declined in popularity), or modified to fit a number of other engine/model year combinations. Consolidation of part numbers is a common method of reducing inventories, but it can make things confusing when a company has original equipment, traditional aftermarket, and performance product lines.

The conversion from the old to the new numbering system has also affected service parts. As an example, electric fuel pump number P-6230 is now listed as part no. 12-802. These all-numerical listings were phased into all parts of the Holley line, so even though the alphanumeric designations disappeared years ago, some references to them can still be found in old literature and parts boxes. In the current part-numbering system, two or three separate number groupings are typical. For example, consider part 0-4781-2. The prefix indicates a general parts grouping by either type of application or type of component. The second grouping indicates the basic part description. The suffix grouping indicates a revision number. If there is no revision, the part number will consist of only the first two group-

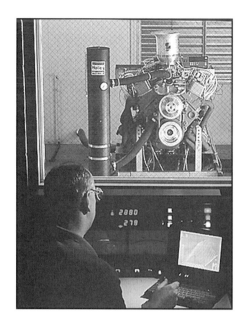

Dynamometer testing of carburetors enables Holley engineers to verify proper air/fuel ratios throughout a wide range of operating conditions. It's one of the reasons that little if any adjustment is needed when a new Holley carburetor is installed.

Chevrolet wasn't the only Detroit automaker to offer three Holley two-barrels as original equipment. Chrysler offered a "Six-Pack" on both big- and small-block engines back in the "good old days."

ings. So, in this example, we have an aftermarket replacement carburetor with the basic part designation of 4781, and this is the second revised edition of this part. (Revisions usually indicate minor design changes, such as alterations of the throttle linkage or other small changes that do not significantly alter the basic operation of the component.)

Some Holley part numbers also include a "size" or operational description code. The most obvious example of this designation is the numbering of main metering jets and power valves. Most of these parts bear a number consisting of two groupings. For example, all Holley standard main jets have the basic part number 122- (formerly 22BP-40-, "BP" indicating the part was bubble packed) which is followed by a suffix group of numbers that refers to the jet number.

Occasionally, throughout the rest of this book, references will be made to product part numbers having an XX suffix. In such cases, the part number refers only to the basic designation and the "XX" symbol indicates several specific parts are available under this basic part designation.

PREFIX	DESCRIPTION
0-	Aftermarket replacement carburetor
1-	Standard replacement (service carburetor)
2-	Economaster line
3-	PEP (rebuild) kits
12-	Fuel pumps and regulators (performance and standard)
16-	Fuel pumps (Standard replacement) Economaster
17-	Adapters and spacers
20-	Small parts
37-	Carburetor repair kits (Renew and Trick kits)
45-	Choke parts
108-	Gaskets
162-	Filters
240-	Breathers
241-	Valve covers
300-	Intake manifolds and cylinder heads
301-	Manifold installation kit
500-	Fuel Injection
800-	Ignition

BASIC CARBURETOR NUMBERING

MODEL SERIES	DESCRIPTION
2000	Standard two-barrel (example 2100, 2300)
4000	Standard four-barrel (4150, 4160, 4360, 4500)
5000	Staged two-barrel (5200, 5210, 5220) (On 5200, the last two digits refer to the original-equipment customer — 5200 Ford, 5210 GM, 5220 Chrysler)

Super Tuning and Modifying Holley Carburetors **13**

Super Tuning and Modifying HOLLEY CARBURETORS
Selecting the Right Holley Carb

Prior to the advent of the four-barrel carburetor in 1952, the only effective means of increasing the breathing capacity of an engine was through conversion from single to multiple carburetion. Early attempts at high-performance V-8 carburetion typically consisted of a specially made aftermarket intake manifold and two, three, four, six or eight Stromberg "97" carburetors. Airflow ratings were not generally available, but back in the days when the Ford flathead V-8 was king of the highway, this was a moot point. There were only a few carburetors that would fit the Stromberg 97 three-bolt mounting flange, which was commonly used by high-performance intake manifold manufacturers.

Later, efforts at multiple carburetion used the larger Rochester 2GC, which had a four-bolt mounting flange. In the late 1950s, even General Motors got into the act, with triple two-barrels on the Oldsmobile "J-2," Pontiac and Chevrolet 348. But the most notable (although ill-fated) development in multiple two-barrel manifolds was the handiwork of a company by the name of Man-A-Fre. By positioning a Rochester two-barrel directly over each pair of cylinder head ports, Man-A-Fre offered the advantages of direct-port induction. In theory, positioning a carburetor barrel directly over each port afforded a significant power increase. In practice, a plenum area between the carburetors and the manifold runners solves a number of problems and generally produces more power. When dual four-barrel arrangements, which provide the same eight carburetor barrels as four two-barrels, came along, the Man-A-Fre manifold galloped off into oblivion, where it continues to serve as a footnote for writers of carburetor books.

When awareness of carburetor airflow capacity became commonplace, bigger was typically equated with better, and "too much" was just barely enough since even the largest available four-barrel was not capable of meeting the air flow requirements of a highly modified V8. Two of the biggest early-model Holley 4150s provided a total airflow capacity of only 800cfm (600cfm four-barrels were not offered until 1958), and many factory dual four-barrel options used even smaller carburetors. It is interesting to note that dual four-barrels first appeared as original equipment in 1956 (Chevrolet), whereas a factory triple two-barrel installation wasn't available until 1957 (Olds and Pontiac).

By comparison, the largest standard-flange Holley four-barrel currently available flows 950cfm. The larger square-flange "Dominators" flow up to 1250cfm. Obviously, with the general availability of such grandiose carburetors, bigger no longer necessarily equals better. In fact, more often than not, a smaller four-barrel is preferable to an extremely large one. While a 600cfm unit with vacuum secondaries may not offer the glamour and panache of a 950 double-pumper, the former is much more appropriate for a mild street engine as it will provide better driveability, greater fuel economy, and crisper throttle response than the latter. It's also considerably less expensive.

Believe it or not, this is where high-performance carburetion was born — on a four-cylinder Ford engine. The Stromberg two-barrel was originally used on the flathead Ford V-8, but it offered a significant performance advantage when adapted to a four-cylinder powerplant.

When two or three Stromberg 97s were bolted onto a custom flathead intake manifold, the airflow race was on. Today, a single Holley four-barrel offers more airflow capacity than four or five old Strombergs.

In the '80s a new trend developed — 200-mph "outlaw" full-bodied cars. As might be expected, the carburetor of choice was a Holley Dominator — a couple of them actually. These cars were the genesis of the Pro Modified class.

For a time it was rumored that the inline Autolite was going to threaten Holley's supremacy in performance carburetion. But only a few Autolite inlines were ever produced, and they never fully proved themselves.

DETERMINING FLOW REQUIREMENTS

Choosing the optimum performance carburetor involves somewhat more of a decision-making process than simply purchasing the largest model your budget allows. The first step is to determine the maximum airflow potentially demanded by the engine that's to receive the new carb. On the surface, this appears to require no more than converting cubic inches (engine size) to cubic feet, multiplying by maximum RPM (to determine the engine cubic foot per minute requirement), and selecting a carburetor that offers a corresponding airflow capacity. Intended usage, engine efficiency, engine operating range and total number of throttle bores must all be taken into consideration.

The basic mathematical formula for relating engine size and RPM to carburetor airflow capacity is shown at right.

This formula assumes 100% volumetric efficiency and does not take into account the pressure differential (drop) at which the carburetor is rated. Although they agree on little else, most carburetor manufacturers have standardized on: a) 3.0 inches of mercury (in/Hg) as the pressure drop

---

CARBURETOR CFM =

$$\frac{\text{ENGINE CI}}{2} \times \frac{\text{MAXIMUM RPM}}{1728}$$

Divide by 2: four-cycle engine intakes once every revolution

Factor to convert cubic inches to cubic feet (12x12x12=1728)

For ease of use, this formula may be modified to:

CARBURETOR CFM =

$$\frac{\text{ENGINE CID} \times \text{MAXIMUM RPM}}{3456}$$

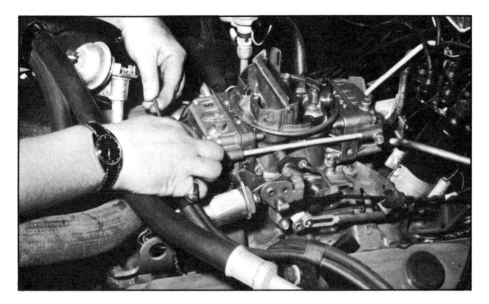

Installation of a carburetor is always easier if the proper part number is chosen at the outset. Holley offers many carbs with the same airflow capacity, but they are calibrated for specific models and years. Linkage and fuel connection provisions also are included for quick and easy installation.

Although dyno testing is a vital part of a full-tilt engine development program, race-track testing is still essential to developing maximum performance.

"On vehicle" testing of Holley products includes sophisticated laboratory analysis as well as thousands of highway miles.

at which two-barrel carburetors are rated, b) 1.5 in/Hg as the pressure drop at which four-barrel carburetors are rated. Since airflow capacity increases (in a non-linear fashion) as pressure drop is raised, the flow ratings of two- and four-barrel carbs can be equated as follows:

> CFM FLOW RATING
> AT 3.0 IN/HG ÷ 1.414 =
> FLOW AT 1.5 IN/HG
>
> CFM FLOW RATING AT
> 1.5 IN/HG X 1.414 =
> FLOW AT 3.0 IN/HG

These differences are based on presumed maximum vacuum attainable under load at wide-open-throttle (w.o.t.). Theoretically, an engine in this situation will not realize an intake manifold vacuum of more than 1.5 in/Hg when equipped with a four-barrel, nor more than 3.0 in/Hg when fitted with a one- or two-barrel. Similarly, an engine that inhales through two four-barrels may never reach a w.o.t., manifold vacuum (under load) of over .75 in/Hg, which explains the use of two 1050cfm carbs on some 330-350 cubic inch race engines. Since 1.5 in/Hg of vacuum will never be attained, the total airflow potential of 2100cfm is equally unobtainable. Simply stated, with any given engine, as total throttle bore and venturi areas increase, maximum obtainable w.o.t. vacuum decreases.

Obviously, carburetor airflow ratings are not intended as absolutes but merely as relative guidelines. Depending on displacement and volumetric efficiency, an engine may flow more or less air through a carburetor of a particular rated capacity. In most instances the manufacturer's rating will be "close enough for government work," so it really doesn't pay to become entangled in pages of calculations. The best bet is to follow the recommendations of the carburetor manufacturer.

However, there are instances where particular engine/chassis combinations are intended for other-than-commonly-encountered applications. In these situations, calculating the airflow requirements of an engine may be advantageous. By way of example, consider a 350 cubic inch powerplant with a maximum engine speed of 8,000 RPM.

$$350 \text{ (CID)} \times 8{,}000 \text{ (RPM)} \div 3456 = 810 \text{CFM}$$

With the 810cfm requirement, an 800cfm carburetor would be ideal, and a Holley double-pumper (#0-4780) would serve nicely for applications such as road racing, street/bracket type drag racing, slalom/gymkhana/ autocross competition, off-road racing and selected short-track oval competition. It should be noted that no adjustment has been made for volumetric efficiency and, obviously, few engines ever reach 100% v.e. Most operate in the 70%-95% v.e. range. Therefore, the 810cfm figure cited in the above example is strictly theoretical, and should be modified:

VE %	CFM REQUIRED
70	567
75	608
80	648
85	689
90	729
95	770
100	810

Volumetric efficiency is not constant throughout the RPM range and is usually highest at the engine speed where maximum torque is being produced. For general reference use, 75% v.e. can be used for low-performance engines although many "smog motors" produced during the 1970s and early 1980s never make it much past a 70% v.e. An 85% v.e. applies to most high-performance engines and a 95% v.e. is appropriate for fully modified racing engines. The

The old Super Stock cars of the early '70s showed how well a stock Holley could perform. Many of these vehicles ran as well as others did with injectors.

Back in the days before the federal government decided it wanted to be in the automobile business, Detroit produced a number of engines fitted with multiple carbs. This big-block Ford is topped by three Holley Model 2300 two-barrels.

English translation of all this is that, instead of consuming 350 cubic inches of air and gas every two revolutions, a 350 engine will use:

262ci operating at 75% v.e.
297ci operating at 85% v.e.
332ci operating at 95% v.e.

In spite of the best efforts of engine builders and bench racers, the laws of physics prevent these percentages from changing very dramatically. Intake manifold efficiency, valve and port size, camshaft timing and exhaust manifold configuration are a few of the more readily identifiable factors affecting the volume of intake charge that will reach a cylinder prior to the power stroke. The low pressure created by a piston moving downward in a cylinder (during the intake stroke) is not sufficient to draw in 100% of the volume required to completely fill that cylinder (with the piston at the bottom of its travel). Therefore, the effect of inertia is needed to keep the incoming air/fuel mixture flowing after the piston has started moving upward (during the initial stage of the compression stroke). The inertia or "ram" effect increases with RPM, which is one of the reasons that internal combustion engines produce maximum horsepower in the upper RPM ranges.

Correlating airflow and volumetric efficiency theories to real life may seem problematical because theory and practice frequently do not coincide. When Detroit was producing high-performance carbureted engines in the 1960s, the original equipment carburetors, in retrospect, appear to have been too large. Chevrolet used an 850cfm carb on the 427 Corvette and a 780 on the 302ci Z-28 Camaro; Ford used a 715 on the 289 and two 540cfm units on the 406 engine. Plugging these engine size and carburetor cfm units into standard flow-requirement equations, the answer "does not compute." It would appear that the factories were expecting 8500+ RPM out of the smaller engines and 7500+ from the larger ones. Most stock engines would encounter valve float or serious internal damage before reaching these RPM levels, and nobody knows this better than the people who produce those engines. But 427 Corvettes, Z-28 Camaros, 289 Mustangs and the like were frequently entered in sanctioned competition where the original carburetor had to be retained. Use of a carb that was too large for street use also served to insure that race engines were not undercarbureted. By comparison, engine options that were inappropriate for competition use were invariably fitted with much smaller carburetors. The 390hp/427 and 350hp/327 Corvettes rolled off the assembly line with a 585cfm Holley, and the 428 Mustang used either a 675 or 735cfm unit. (As further evidence of the airflow formula's accuracy, the larger carburetor was used only in 1968 and 1969 when the 428 Cobra Jets were raced extensively in Super Stock drag competition.)

VACUUM OR MECHANICAL SECONDARIES

In the real world, it would appear that optimum carburetor size can in fact be calculated accurately — so long as realistic RPM and v.e. figures are used, over- or under-carbureting

Unlimited race-type carburetion really came of age during the early days of Pro Stock drag racing. Like the cars themselves, Pro Stock carburetor technology continued to evolve and is one of the reasons these cars now run six-second quarter-mile times at over 200 mph.

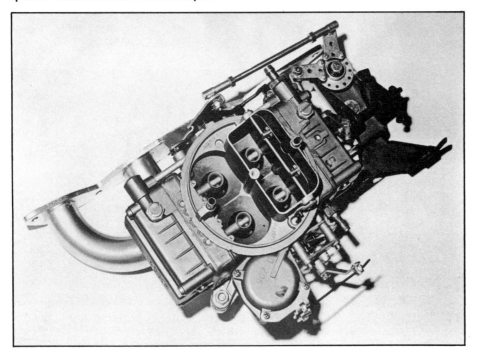

There is even a Model 4160 for four-cylinder applications. Part number 0-6299, the "Mini-Muscle" carb, is rated at 390cfm and features a reverse power valve and special float assembly to alleviate vibration problems. This carburetor has been converted to 4150-type bowls and metering block through use of kit 34-6.

should not be a problem. The only other consideration is secondary throttle activation method on four-barrel carburetors. Vacuum control offers potentially smoother operation, and minimizes the effect of a carburetor that is too large, because it has the ability to "size" the carburetor to the needs of the engine. That statement assumes the vacuum actuating mechanism is properly tuned and opens the secondary throttles at the appropriate rate, thereby allowing maximum air velocity through the carburetor to be maintained at all operating RPM levels.

With vacuum actuation, the secondary throttle plates are actually pulled open by air flowing through the primaries. If a carburetor is too large for a particular engine, sufficient venturi vacuum (not manifold vacuum) to fully open the secondaries may never be generated. Therefore, a 780cfm carburetor may never flow more than 650cfm, if that's all the engine requires.

Conversely, mechanical secondaries offer the advantage of allowing the driver to control precisely when the secondary throttles are opened. This is especially important in oval track and road racing where performance "coming out of the corners" is critical. In drag racing, positive secondary throttle control is also advantageous, so use of vacuum secondary carburetors is uncommon except in Stock and Super Stock classes where rules dictate retention of the original equipment carb.

Vacuum control is typically advised for street, R.V. and recreational boat use, while mechanically operated secondaries are found in virtually all competition applications, the notable exception being off-road racing. Due to the inconsistency of the terrain and lack of gas pedal control that arises from the driver bouncing all over the passenger compartment, the smooth application of power is most important. If the secondary throttles are fully opened when the gas pedal is pushed to the floor, excessive wheelspin can result, causing the rear tires to dig a hole rather than pushing the vehicle forward. Vacuum secondaries alleviate this problem by opening gradually as demanded by airflow. Information pertaining to modification of opening rate is included in later chapters.

CARB-MANIFOLD-CAM INTERRELATIONSHIPS

Holley offers a wide assortment of performance carburetors for virtually any application. Once an air flow capacity is decided upon, choosing a proper part number is rarely a problem. But there are occasions when "the right" carb simply doesn't afford the performance levels desired.

Regardless of horsepower figures produced on a dyno, or computed through engineering theory or mathematical formulas, the bottom line of the performance ledger concerns the ability of a carburetor to function in a real world environment. Integral parts of this environment are the intake manifold and camshaft. In many cases a carburetor is expected to compensate for inadequacies in the intake system. The novice tuneup artist may, in such cases, turn to a carburetor of different airflow capacity to alleviate some undesirable operational characteristics, but such a solution is always a weak compromise.

A carburetor cannot compensate for the inherently bad low-RPM characteristics that are typical of long-duration cams and single-plane manifolds. Consider the case of a typical performance enthusiast with more money than brains. (Sound like anyone you know?) Determined to impress his friends, he's really doing little more than entertaining himself when he installs a long-duration camshaft, a competition-type, single-plane manifold, an 850cfm double-pumper carb and 2-inch diameter racing headers on his 350 Chevrolet engine. After spending a small fortune and hours of time installing his newly acquired parts, the car delivers neither the neck-snapping performance he expected nor any sort of reasonable gas mileage.

When the staccato idle no longer compensates for poor performance, the first corrective measure he takes is to install a smaller carburetor in a futile attempt to restore low-speed and mid-range response. However, the combination of a big-port intake manifold, large-diameter headers and a super-lumpy camshaft will not allow this, or any carburetor, to produce normal fuel delivery. Reduced vacuum at idle (created by long cam overlap and weak low-RPM exhaust scavenging) will delay activation of fuel flow through the main nozzles, resulting in an off-idle stumble. The problem can be rectified somewhat by increasing accelerator pump discharge volume or by enriching the idle circuit a great deal, but both of these "fixes" are Band-Aids when major surgery is required. A much more satisfactory approach would be to switch to a shorter duration camshaft and a dual-plane street-type manifold. (If you're convinced you must run a ram manifold on a street engine, keep cam duration relatively short — 220° [at .050-inch lift] or less — and use two 390cfm carburetors on a Holley Pro Dominator or similar manifold.

OTHER CONSIDERATIONS

When a carburetor is purchased, the buying decision should be made based on an honest appraisal of the equipment with which the carb must interact and the RPM level at which the engine will realistically operate. Assuming that the cam, manifold and headers in the example above had already been installed, a 600cfm carb with vacuum secondaries would have provided better driveability than the 850 double-pumper. (Even on a race engine, too much carburetor airflow capacity can reduce performance to below optimum levels.) But there is more to carburetor selection than mere size consideration. Certain economy/emissions carburetors are fitted with booster venturis that are very sensitive to changes in the manifold conditions. When these carburetors are used in conjunction with a long-duration camshaft, the pulses (in the intake manifold) generated by excessive valve overlap can degrade fuel metering capabilities. A single-plane manifold with a large, open plenum intensities such problems.

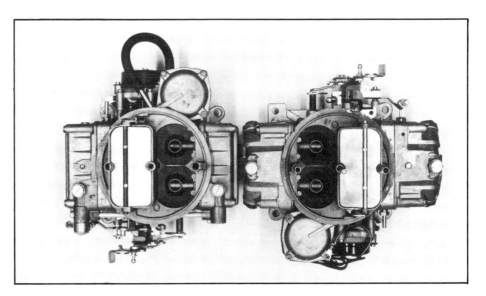

On the left is a 390cfm Model 4160, while the carb on the right is a 600cfm version. Note the difference in venturi size.

This experimental four-barrel contains "Economaster" boosters in the primary venturis. These boosters are designed to provide improved atomization but are not available in production four-barrels.

For underhood pizzazz, it's hard to beat a Dominator, hence the existence of a 750cfm street version. However, use of this carburetor requires a manifold specifically designed to accept a Dominator or an adapter.

Super Tuning and Modifying HOLLEY CARBURETORS
Basic Functioning

If anyone ever required proof that a little knowledge is a dangerous thing, he or she need only spend some time with a few shade-tree mechanics or "hero" racers. Eager to prove their mastery of automotive technology, many of these self-proclaimed wizards attempt to diagnose and correct engine problems without gaining a full understanding of the functioning of the components that allow the miracle of internal combustion to become reality. If you have read any of the preceding chapters, or even the front cover, you have no doubt already figured where this paragraph is leading — right to the carburetor that sits atop many a high-performance engine.

Part of the reason that carburetors in general, and Holleys in particular, are reputed to be complicated has to do with the danger associated with less-than-adequate amounts of knowledge. This chapter is intended to lead you safely out of the danger zone.

Except for a few eccentric designs, all modern automotive carburetors are very similar in concept. Moreover, they are identical in basic function; differences lie in the methods used to accomplish those functions. Ironically, the Holley 4150/4160 four-barrel, with its intimidating reputation, should actually be the easiest of all multi-barrel carburetors to comprehend since the modular construction lends itself to investigation and discovery. But rather than immediately exploring the operation of a Holley four-barrel, building the bridge to understanding should begin with an examination of carburetion principles.

PRIMARY METERING SYSTEMS

The basic purposes of a carburetor are fuel atomization, air/fuel mixing, and airflow regulation, which govern engine speed. These concepts can be understood most easily by exploring the theory of individual aspects of carburetion and then relating the theory to functional hardware as incorporated in Holley's family of high performance carburetors.

In order for gasoline to burn properly within a combustion chamber, it must be rendered into an atomized state and mixed with some amount of air. Air is necessary to provide oxygen. Prior to the days of exhaust emissions consciousness, 14.7 parts of air (by weight) to one part of gasoline was thought to be the optimum air/fuel ratio for part throttle cruise conditions (current technology has enabled engines to run as lean as 17:1 or 18:1). Under full-throttle, heavy-load operation a richer mixture (around 12:1 or 13:1) is generally desired.

When fuel exits a carburetor's discharge nozzle under proper conditions, it has been partially atomized (converted to a spray of fine droplets). It is not until fuel reaches the low-pressure/high vacuum area beneath the throttle plate that it becomes thoroughly atomized. Strong manifold vacuum is a necessity to "draw" fuel out of the main discharge nozzle as well as to create the velocity through the carburetor venturi to enable atomization, leading to good vaporization and thorough fuel combustion.

If an engine were required to operate under a constant load, at a constant RPM, the throttle opening would be permanently set and atomization conditions would be ideal since manifold vacuum and air velocity (through the carburetor) would always be the same. The carburetor for such an engine would need only one fuel metering circuit, the one that is commonly referred to as the main system. Mixture (air/fuel ratio) would be controlled by jet size, and fuel would be discharged through a nozzle located in the low pressure portion of the venturi.

Over 170 years ago, an Italian scientist who had no interest whatsoever in automobiles discovered that when air moves through a tube with a double-taper (diameter changes from large to small and back to large) the velocity is highest and pressure lowest in the area of smallest diameter. This principle, named for its discoverer, Venturi, forms the basis for virtually all carburetor design. Like a vacuum cleaner, the low pressure condition found in a venturi draws fuel out of a reservoir (float bowl) through the discharge nozzle.

With low pressure in the venturi acting in the same manner as a suction pump, there is no reason for fuel to exit the discharge nozzle in anything other than a liquid state. Just stick an "Oklahoma Credit Card" (siphon tube) into a gas tank and apply suction to the other end. Rather than atomized petroleum distillates, you'll get a mouthful of pure high-octane gasoline. Mere suction serves only to initiate flow; it provides no means of converting liquid gasoline to an atomized mist. Atomization results when a stream of liquid is introduced into a column of fast moving air, but without an additional step, many of the gasoline droplets are too large to be effectively atomized for optimum combustion efficiency. The additional step needed in the liquid-to-mist breakdown process

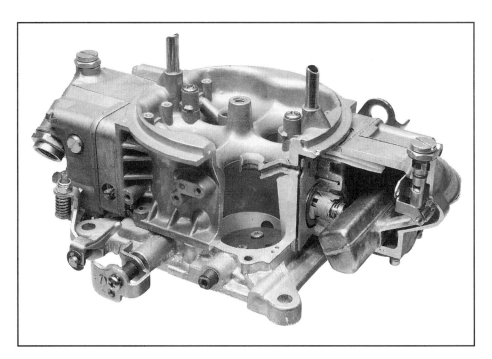

This Model 4150 race carburetor, with dual accelerator pumps, adjustable secondary idle circuit, and dual-feed float bowls may seem complicated, but once the basic operation is understood, it is one of the easiest carburetors to custom tailor for specific applications.

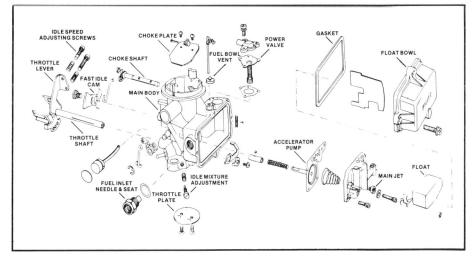

is accomplished between the fuel bowl and discharge nozzle and is known as emulsification, or pre-atomization.

According to dictionary definition, an emulsion is "a mixture of mutually insoluble liquids in which one is dispersed in droplets throughout the other." If one were to take quantities of vinegar and vegetable oil, pour them in a bottle and shake vigorously, the end product would be an emulsion which is commonly known as a salad dressing. If the emulsion is allowed to stand, the two liquids will soon separate and the lighter substance will rest entirely on top of the heavier one.

This exploded view of a typical one-barrel carb illustrates basic carburetor construction.

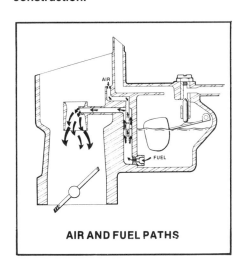

AIR AND FUEL PATHS

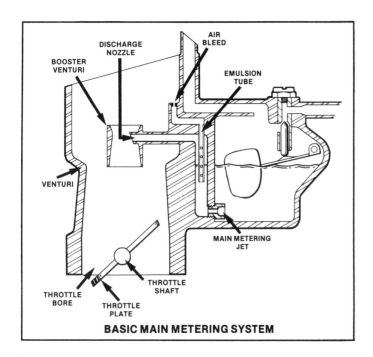

BASIC MAIN METERING SYSTEM

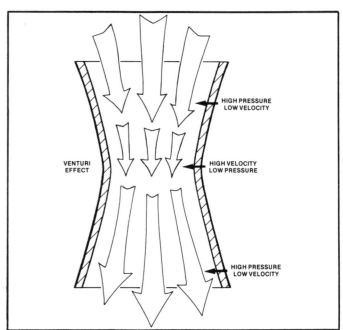

The main metering system includes: main jet, emulsion tube, air bleed, discharge nozzle, and booster venturi.

In the world of carburetion, an emulsion refers specifically to droplets of fuel interspersed with a volume of air. The device most instrumental in emulsification is known as an air bleed. With an air passageway leading from the bleed to a point between the discharge nozzle and the float bowl, the same suction (low pressure) force that causes fuel flow now draws air, in addition to fuel, into the delivery tubes. The effect is similar to the experience of trying to draw up a drink's final drops from the bottom of a glass with a straw. Much of the liquid resting alongside an ice cube is drawn up the straw with some amount of air. In some instances, the liquid breaks down into smaller droplets as it travels up the straw. So in a sense, there is no difference between an emulsified Pina Colada and the air/fuel mixture that exits at the main discharge nozzle. However, no one has yet been successful in operating a high-performance engine on Pina Colada fuel.

However, the function of an air bleed is not solely to emulsify fuel. It also exerts control over fuel flow by "bleeding off" some of the suction force or "signal" which exists at the discharge nozzle. By varying the size of the bleed, the amount of suction (vacuum) required to initiate fuel flow can be altered. As bleed size is enlarged, the amount of vacuum (and hence the engine RPM) necessary to initiate fuel flow is increased. Conversely, a reduction in bleed size reduces vacuum requirements.

Consideration of extreme examples will simplify the understanding of this concept. If the bleed could be infinitely enlarged, the point would be reached where a suction pulse was never transmitted to the fuel channel. In essence, the entire vacuum "signal" would be bled off. Total elimination of the bleed, on the other hand, would allow transmission of full signal strength to the fuel reservoir.

The air bleeds within production carburetors are precisely measured restrictions, sized so that fuel flow is started at specific vacuum levels. (It should be noted that this discussion pertains to venturi vacuum, not manifold vacuum. Venturi vacuum increases with airflow and will therefore be greater at wide open throttle than at idle or part throttle. This, of

Carburetor functioning is based on the Venturi Effect. As air moves through a venturi, velocity increases and pressure decreases (drops). The pressure drop is used to "draw" fuel into the airstream. Main fuel metering is accomplished by the pressure drop inside of the booster venturi, thus drawing fuel from the nozzle located inside the booster venturi.

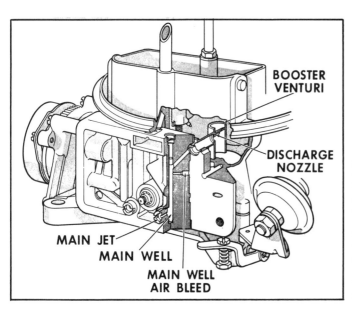

Main metering systems are typically straightforward in design and operation. Fuel enters the main jet, is drawn up the well, emulsified with bleed air, and dispersed into the incoming airstream through the booster venturi nozzle.

Super Tuning and Modifying Holley Carburetors

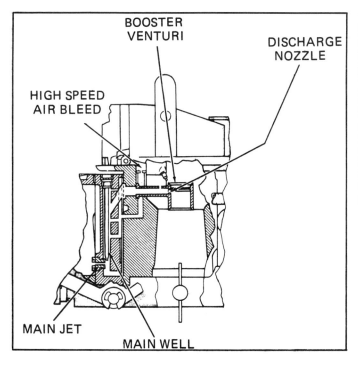

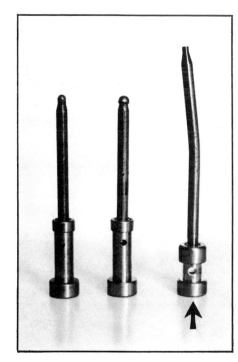

The function and importance of the air bleeds often are overlooked. The size (area) of the bleed controls circuit "startup" timing. Sophisticated flow equipment is required to coordinate bleed size and circuit timing of the idle and main fuel circuits to gain efficient overall performance.

There are several styles of emulsifier tube, and sometimes what appears to be such a device is not. At right is the idle supply tube used in Model 4500 carburetors equipped with an intermediate circuit.

course, is in direct contrast to manifold vacuum which decreases as the throttle is pushed toward wide open position.) Without the sophisticated equipment required to perform fuel flow analysis, it is impossible to accurately assess the results of bleed size alterations. Trial-and-error testing will obviously define the effects of air bleed diameter alteration, but unless a carburetor is fitted with removable bleeds, reducing orifice size is difficult. This is the reason most carburetor engineers advise against such modifications.

It should be somewhat obvious that bleed size also affects air/fuel ratios. A large bleed will admit more air than a small one and, consequently, a leaner mixture results as bleed orifice diameter is increased.

At this juncture, it should be obvious to anyone who has looked down the throat of a Holley four-barrel that something more sophisticated than a simple discharge tube mounted in a low-pressure area is required for satisfactory performance on a modern automotive engine. With a mere tube protruding into the venturi, serving as the discharge nozzle, inordinately high airflow is required to develop a sufficiently strong signal. A venturi simply does not provide a large enough pressure drop to initiate fuel flow at lower air velocities. In essence, the pressure drop across a venturi needs to be amplified. The device used to perform this function is commonly called a boost or booster venturi.

Use of a booster venturi, as opposed to reducing the diameter of the main venturi, offers a number of advantages, although they pertain primarily to production considerations. Rather than requiring a different main body casting (with a venturi of a particular diameter) to "fine tune" a carburetor for flow capacity and signal strength, one casting can be employed, and the shape and size of the booster altered, to achieve the desired effect. The cost of changing a relatively simple booster is far less than the expenditure necessary to alter an intricate main body casting. (This is only viable over a relatively small range.) As an example, part number Holley 0-3310-1 has the same venturi and throttle bore dimensions as the 0-3310-2, -3 and -4, yet it flows 780cfm, as opposed to 750cfm. The 30cfm difference in flow capacity is caused by a more restrictive booster, which in turn offers a stronger signal.

Another advantage of the booster concept is that vacuum can be intensified in the area that immediately surrounds the discharge nozzle. By varying booster shape, it is possible to increase signal strength while minimizing loss of airflow capacity. One of the reasons that the 4150/4160 series Holleys have

A typical metering block with selected areas milled away to reveal internal fuel passages. Fuel enters main well through the jet, then travels up toward the angled passage that leads to a main discharge nozzle in the main body. The tube that extends down into the main well admits bleed air to emulsify the fuel.

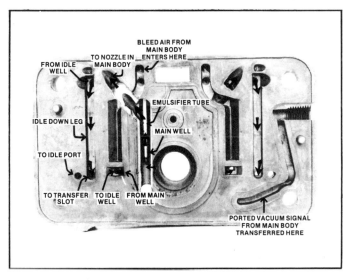

Super Tuning and Modifying Holley Carburetors

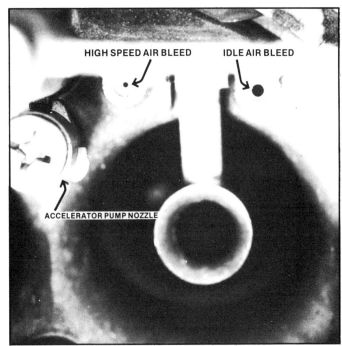

A view down the primary venturi of a Model 4150/4160 Holley. Air bleeds are precisely drilled brass restrictions that control delivery "timing" as well as air/fuel ratio. Idle air bleed is typically larger in size than the high-speed bleed.

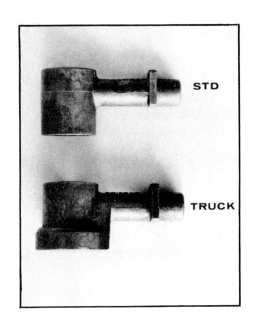

On top is a standard booster that provides a relatively strong signal. It is more restrictive than the "down-leg" booster, but less so than the "truck" design. Signal strength increases with amount of flow restriction (i.e., pressure drop).

Although they are not replaceable without special equipment, several booster venturis are used by Holley. At right is the vane-type booster used in some Economaster models, and at left is the bell-shaped "truck" version (originally used in a truck carburetor, hence the name).

proven so versatile is that Holley has been able to use several booster designs without changing the basic carburetor casting. Of the current Holley booster designs, the down-leg version is least restrictive, followed respectively by the right-angle and bell-shaped or "truck" configuration. As might be expected, the truck booster is known for its strong signal.

However, these boosters, like many other designs, are limited in their ability to distribute fuel evenly around the throttle bore. When poor manifold designs compounded this problem, Holley engineers, working with original-equipment carburetors, either added a tab or milled off a portion of the booster's trailing edge. These modifications "shaped" the low-pressure area such that exiting fuel was pulled into a more desirable distribution pattern when a particular carburetor was installed on the manifold for which it was originally designed.

With the changing priorities over the years, fuel distribution became especially important. In 1976, Holley released the Economaster line of carburetors and included a revamped Model 2300 (two-barrel) in the initial offering. Externally, this carburetor looks like any other two-barrel in the 2300 family, but it contains a totally unique vane-type booster. Eight small arms, connecting an inner circle to the outer diameter, create eight corresponding low-pressure areas on their undersides. This not only enhances fuel distribution but also improves atomization since the localized low-pressure areas (beneath each vane) cause the fuel to be broken into smaller droplets. This design is, unfortunately, highly restrictive to airflow and therefore not suited for high-performance use.

Another radical departure from Holley's traditional booster design is the annular-discharge booster released in early 1980. Like its vane-type counterpart, this booster offers improved distribution and atomization, but not at the expense of airflow. Instead of a single discharge nozzle, as found in conventional boosters, fuel is channeled through the body of the booster and enters the air stream through eight small holes distributed about the inner diameter. The annular-discharge booster provides an exceptionally strong signal, making it ideal for racing applications where long duration camshafts cause a reduction in manifold vacuum at lower engine speeds.

Since each individual booster design creates a vacuum signal of different intensity, it can be seen that fuel metering requirements and air bleed dimensions can vary significantly among carburetors, even among those with identical throttle bore and venturi diameters. By dealing with these variables, carburetor engineers and modifiers can

There are different versions of the "down-leg" booster. The cutout on the piece at lower left is used to alter fuel distribution; the tube in the booster at right has a similar effect. "Down-Leg" configuration is the least restrictive of traditional designs.

Since the main body differs in design, the Model 4500 has a special booster design. At left is a standard booster, at right is a special model designed to control fuel problems in individual runner (IR) race applications.

Holley's annular discharge booster develops a strong signal without being highly restrictive. It is available in certain Model 4150 and 4500 race carburetors.

create an almost infinite variety of fuel delivery characteristics.

Although many modified carburetors feature altered booster venturis and air bleeds, these items were never designed to accommodate casual modifications. Misdirected enthusiasm has ruined more than one perfectly functioning carburetor. However, the metering jets can be removed and replaced by anyone who understands the operation of a screwdriver. The jet is therefore the preferred method of altering main system fuel flow to suit a specific requirement.

A carburetor jet is nothing more than a metering device that controls the flow of fuel from the fuel bowl into the main circuit. Once metered by the jet, fuel travels through the main well, where it is emulsified prior to flowing into a connecting channel that leads to the discharge nozzle. In all Holley Model 2300, 4150, 4160, 4165, 4175 and 4500 carburetors, the main wells are vertical passages cored into the metering block. About 2/3 of the way up the passage, air from the bleed is introduced so the fuel is emulsified before it leaves the metering block and enters the main body.

In model 4150/4160, 4165/4175 and 4500 carburetors, Holley generally introduces bleed air by one, or sometimes a combination, of the following methods:

1) A brass tube with a number of small holes drilled through the wall is inserted into the main well. Air is drawn in through the bleed in the main body and through a passage that leads to an external channel on the metering block. A hole in the channel connects to the inside of the brass tube, which in turn allows air to be drawn out of the tube (through the small holes) into the main well.

2) Air is drawn through the bleed in the main body and through a passage into the external channel in the metering block. Holes drilled from the channel directly into the main well admit the bleed air. In either case it is the low pressure (pressure drop) created by the booster venturi that causes air to be drawn in through the bleed and fuel to be drawn through the jet. And, as long as air and fuel are mixed in proper proportion, the main metering circuit is performing as it should.

In Holley model 4010 and 4011 carburetors, the main wells are cast into the carburetor body and the emulsion tubes are attached to the booster venturi cluster (which also contains the accelerator pump discharge nozzle). Positioned inside each emulsion tube is an idle tube through which fuel flows on its way to the idle discharge port. A precisely sized orifice at the base of the idle tube serves as an idle feed restriction which meters fuel into the idle circuit.

POWER ENRICHMENT

So long as the signal from the booster remains constant or increases in intensity (from a corresponding increase in airflow), fuel flow will be sufficient to maintain the air/fuel ratio that is an inherent part of a carburetor's design. But what

The most basic of all carburetor recalibrations is the simple jet change, requiring only a screwdriver.

Holley has used several power valve designs. The four-hole model (left) is too restrictive for some applications, so the six-hole model was developed. The "picture window" design, with even better flow characteristics became the standard performance unit. Numbers stamped on the flat surface refer to the opening point.

happens when signal strength drops? Obviously, a leaner air/fuel mixture will result; as vacuum decreases, the amount of fuel drawn through a fixed orifice (the jet) also decreases. Whenever the throttle plates are opened rapidly, engine vacuum drops momentarily, as does the vacuum signal that causes fuel flow. Consequently, the mixture grows leaner at precisely the time when the engine demands additional richness — such as during acceleration when the load placed on the engine is heavier than under cruise conditions.

Engine vacuum also drops during periods of extended high load, as when climbing hills or towing a trailer, so a carburetor must make provision for supplying additional fuel for all heavy-load conditions, irrespective of duration. Enrichment for high-load conditions is accomplished by either a "power" or "economizer" circuit.

In days of old (and occasionally in days of new), carburetors were said to have an "economizer valve." This nomenclature indicates a rather negative orientation since it describes the operation of a device when it is not functioning. The name is undoubtedly the product of a brilliant marketing mind who was attempting to justify the added expense of another fuel circuit to a stingy accounting department. (Being given more to extravagance than frugality, I'll approach this subject from a positive angle, with the emphasis on power rather than economy.)

As used in a Holley four-barrel (and some other makes and models), a power valve consists of a spring-loaded rubber diaphragm and a valve-seat assembly encapsulated in an aluminum body which is threaded to allow installation in a metering block or (in the case of model 4010/4011) in the main body.

When manifold vacuum presented to the back side of the diaphragm is sufficient to overcome spring pressure, the valve closes, preventing fuel flow. Conversely, as vacuum drops below the point where spring pressure is overcome, the valve opens and additional fuel flows into the main well. The versatility of Holley carburetors is due, in part, to a broad selection of power valves, with opening points ranging from 2.5 to 10.5 inches of vacuum. Each valve is stamped with a number corresponding to the opening point. (An "85" valve would open when manifold vacuum dropped below 8.5 inches.) A manufacturing code consisting of a letter and a number may also be stamped on each power valve and pertains to the month and year of manufacture. (B9 indicates February, 1989; K9 November, 1989.)

In the early 1970s when the popularity of recreational vehicles surged, Holley released a two-stage power valve. With the opening of the first stage set to occur relatively high (10.5 or 12.5 inches of Mercury), the two-stage valve offers vehicles with low power-to-weight ratios the potential for greater fuel economy, along with improved driveability under moderate load conditions. Partial enrichment, occurring at the 10.5 or 12.5 in/Hg level, (with full enrichment when manifold vacuum drops to 6.0 in/Hg or lower, depending on the specific valve) allows jetting to be somewhat leaner than would be possible with a conventional single-stage valve (because additional fuel is available from the power valve circuit for operation under light-to-medium load). Were a 10.5-inch single-stage valve used rather than the two-stage model, economy would not be as great since full, rather than partial, enrichment would occur when the valve opened. (Engines used to propel campers, motor homes and other heavily loaded vehicles frequently operate at or below 10.5 inches for long periods of time, especially at higher elevations. Since these are not maximum-load conditions, full enrichment is not warranted.)

However, with a single-stage valve set to open at 6.5 inches, the lean jetting that is a practical companion to a two-stage valve would cause surging and stumbling under moderate load. Hence a mixture that is too rich for cruise conditions would be required so that smooth engine operation would be possible during acceleration and on upgrades. In essence, the two-stage valve optimizes fuel economy by tailoring air/fuel ratios more precisely to engine requirements. Such precision is generally not required on passenger cars (except those towing trailers) since vehicle weight is not sufficient to pull manifold vacuum below 10.5 inches during light-load operation. Exceptions to this rule have unfortunately become more frequent with the development of more stringent emissions and fuel economy standards.

Implementation of the power circuit is quite straightforward. Fuel from the float bowl enters through the center of the power valve and exits through openings in the valve housing, adjacent to the threaded area. Two restrictions, known as PVCR's (power valve channel restrictions), are visible when the power valve is removed. They meter the flow of fuel into the main well. It is the PVCR diameter, NOT the openings in the power valve, that controls the amount of fuel admitted through the circuit, except in applications where a two-stage valve is installed in a carburetor that has a PVCR of .060-inch or greater. Two-stage valves are restrictive to fuel flow, so they are not recommended for general high-performance use.

Currently available Holley power valves are of the "picture window" configuration with two rectangular slots used as fuel exit paths. Earlier designs employed four or six holes, rather than slots, and offered less flow potential. A late-model valve may be installed in place of an earlier version but requires a different gasket (supplied with the valve). Part number 125-XX, where the last two digits refer to the manifold vacuum at which activation occurs, is available with opening points ranging from 2.5 inches to 10.5 inches of mercury (e.g., 125-65 opens at 6.5 inches and 125-105 opens at 10.5 inches). Part number 125-XXX is essentially the same valve, but lacks a piloting shoulder for the valve stem. Power valves with a three-digit suffix offer greater flow potential and are required only with extremely large PVCR's in the vicinity of .090-inch.

One exception to the three-digit suffix scheme is the 10.5 in/Hg power valve. The suffix for the standard valve has three digits and the suffix for the high flow valve has four digits. This is obviously necessitated by the numerical truth that three digits are required to represent a value with two digits on one side of a decimal point and one digit on the other.

When using "picture window" power valves, only a round gasket should be used. Gaskets with three small tabs on the inside diameter were used for centering the old-style power valves with four or six holes drilled in their shanks. Since a shoulder is present on the 125-XX valves, no centering tabs are required. Should either power valve type be used with the incorrect gasket, faulty operation can result, as the valve may not seat properly.

Once through the power valve and PVCR's, power circuit fuel enters the main well and follows the same path as fuel metered through the main jet. Power valve operation is controlled by a vacuum signal that is presented to the diaphragm side of the valve. This signal (manifold vacuum) is channeled from the manifold plenum area into the power valve well through a hole drilled in the base plate and main body. Although it rarely happens, a warped metering block or main body, or a damaged gasket, can

Metering of fuel through the power enrichment circuit is critical since a leanout (due to power valve malfunction) under heavy load can cause engine damage. Holley power valves are checked on this machine to ensure the accuracy of opening-point calibration.

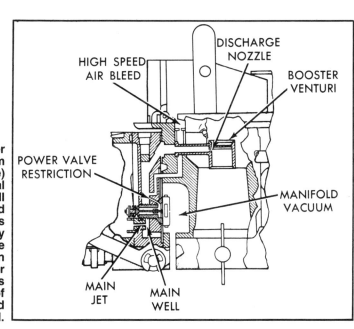

The power enrichment system (power valve) admits additional fuel to the main well when manifold vacuum declines (high speed/ heavy load). Power valve channel restriction (PVCR) diameter (area) determines the amount of fuel metered into the well.

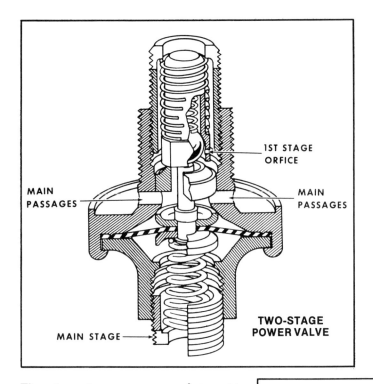

The two-stage power valve was designed for vehicles with low power-to-weight ratios (campers and RVs) to increase light-load cruise economy.

bleed the vacuum signal off, causing the power valve to remain open at all times. However, a more common occurrence (although still rare) is for an engine backfire to rupture the diaphragm. Once again the power valve will remain open, but there is also a possibility that fuel will flow through the damaged diaphragm into the power valve well, and then down into the intake manifold. One might suspect this as the problem if clouds of smoke billow from the tailpipe and no other cause (such as partially closed choke or improper fuel level) can be found. Some Holley carburetors incorporate a ball check in the passage that communicates vacuum signals to the power valve well from the intake manifold. In the event of a backfire, the ball seats and prevents damage to the diaphragm.

ACCELERATOR PUMP

If the same creative mind that contrived the "economizer valve" label had been allowed to run rampant he would have no doubt called the accelerator pump a cruise pump. Fortunately, after the economizer episode, he was trundled off some-

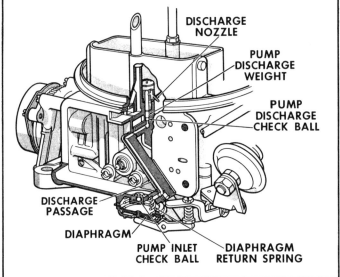

With the surface milled away, the PVCR entry to the main well is visible. In order to reach the channel restrictions, fuel must flow through the center of the power valve and out to the edge.

The accelerator pump housing, located at the bottom of the float bowl, contains a spring-loaded diaphragm. When the throttle is opened, the pump level forces the diaphragm upward (against spring), discharging fuel.

where far removed from carburetor nomenclature. He probably spent his remaining years in a small bleak cubicle in the accounting department. As a result, an accelerator pump has always been known as an accelerator pump.

Its purpose is to provide a quick shot of fuel any time the throttle is opened abruptly. The volume of fuel discharged through the pump nozzle or "squirter" is determined by the number of degrees that the throttle shaft is rotated and the rapidity with which that rotation occurs. Unless movement is fairly abrupt, fuel will be allowed to return to the float bowl rather than being pumped into the air stream. (On the Holley 4150 design the pump is vented back to the float bowl to permit this bypass.)

When a standard or mild high-performance engine is idling, manifold vacuum will generally be between 14 and 17 in/Hg. A correspondingly strong signal maintains satisfactory fuel flow through the idle circuit and the air/fuel ratio will (hopefully) be approximately 13:1. If the throttle plates are opened quickly (as in normal acceleration) several events, which serve to disrupt the mixing of air and fuel, occur concurrently:

1) Manifold vacuum drops to zero, so fuel flow virtually stops.

2) Fuel already in suspension "falls out" when vacuum is suddenly reduced.

3) The new throttle setting requires a significant increase in fuel flow so the main system must be activated. However, without a reasonably strong vacuum signal, flow will not be initiated.

Without an accelerator pump to supply additional fuel during rapid throttle openings, momentary fuel starvation would cause the engine to sputter and/or die. In essence, this circuit "closes a hole" in a carburetor's fuel delivery capability. It should be noted that some small capacity (low airflow) carburetors designed for small displacement engines do not have accelerator pumps. Special booster venturi designs combined with a small throttle bore diameter are sometimes capable of maintaining sufficient vacuum to activate the main system quickly enough to prevent a momentary leanout.

Alteration of pump discharge nozzle diameter, pump capacity, or operating lever geometry influences both pump volume and timing. Most carburetors use a rod-and-lever arrangement for accelerator pump activation. Holley Model 2300 two-barrels and Model 4010, 4011, 4150, 4160, 4165, 4175 and 4500 four-barrels are unique in their employment of a cam, mounted on the throttle lever, in place of the traditional rod-and-lever pump. The cam, separate pump housing, and removable discharge nozzle allow for unparalleled tailoring of the pump circuit.

During the transition from idle to main system metering, fuel discharged through the accelerator pump compensates for the time lag that exists between demand (throttle opening) and main system "start up." This lag is present only when the throttle is moved quickly. During extremely slow opening, airflow increases at such a rate that the transition from idle to main metering is a smooth one, obviating the requirement for additional fuel.

One influencing factor that is not often considered is engine load. Many times an engine will accelerate smoothly when the transmission is in neutral but will stumble badly when called upon to propel a vehicle. Absence of a load enables engine RPM to increase quickly when the throttle is opened; manifold vacuum therefore returns to 14-17 inches without much delay. The combination of high airflow and a strong vacuum signal initiates main system fuel flow rather quickly. Conversely, when a load is present, airflow and vacuum remain low for a considerably greater time period, during which the carburetor merely sits atop the intake manifold with its throttle plates open, not flowing a sufficient amount of fuel to maintain smooth operation.

The duration of the low-signal time period is influenced by many factors including: engine torque, vehicle gearing, vehicle weight, ignition timing, cam timing, carburetor size and engine operating range. For these reasons, it is necessary to tune the accelerator pump for optimum engine operation, for as low-signal (or lag) time increases, so should the volume of fuel discharged at the pump nozzle. In essence, the fuel delivered by the accelerator pump must keep the engine running during throttle transition times when fuel available from other circuits is inadequate.

Carburetor manufacturers use either a piston or diaphragm pump design. The latter type (used on all

Accelerator pump nozzles are available in a variety of styles and sizes. At left is the anti-pullover design employed in 4165/4175 Spread-Bore carburetors. This design prevents fuel from being siphoned out of the accelerator pump circuit. In the center is a tube-type "shooter," and at right is the standard version.

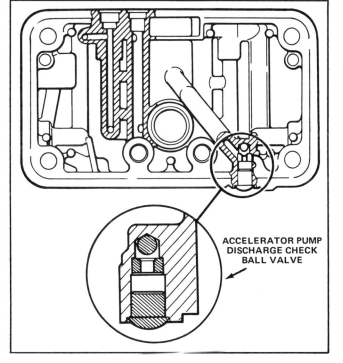

On Model 4165/4175 carburetors, the accelerator pump check valve is located in the base of the metering block. This design necessitates use of an anti-pullover discharge nozzle because all the fuel between the valve and the nozzle is exposed to low pressure at the top of the venturi. At higher airflow rates, fuel will be siphoned into the airstream. Some emissions-calibrated 4150/4160 carbs also use this arrangement.

Super Tuning and Modifying Holley Carburetors **29**

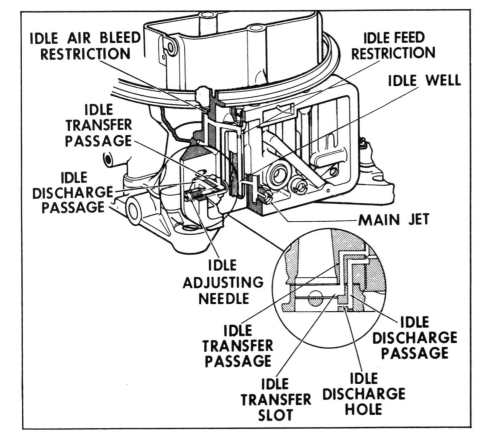

Idle fuel delivery in Holley modular carburetors is determined by fuel restrictions in the metering block and air bleed restrictions in the main body. Curb idle mixture (air-to-fuel emulsion) delivered through the idle discharge hole is controlled by size of idle feed restriction in the metering block and the size of the idle air bleed in the main body. The volume of air/fuel emulsion delivered to the idle discharge hole is controlled by the idle mixture screw. As engine speed builds and throttle plates open wider, the idle mixture begins flowing from the transfer slot, and flow from the idle discharge hole stops.

Holley four-barrels except the 4360) is generally considered a more efficient pump since leakage of fuel out of the pump chamber is virtually non-existent. Piston pumps do not offer a positive seal and when the pump is activated, some amount of fuel will leak out of the chamber between piston cup and the (chamber) cylinder wall. With the passage of time, the plunger (cup) material can shrink and harden, and more fuel will leak out than is discharged at the nozzle. Old carburetors never die, their accelerator pumps merely shrivel.

With a diaphragm pump, leakage is rarely a problem. The only sealing to be done is at the inlet valve and either the steel check ball or rubber "mushroom" valve used by Holley rarely allows leakage. A fairly recent addition to high-performance carburetors, the mushroom valve offers an advantage in that it is closed when at rest. Fuel is therefore discharged from the chamber the instant that the throttle is moved. The steel check ball, on the other hand, is normally in the open position and must be closed (or seated) before fuel is moved to the nozzle. Typically, the clearance between the check ball and the seat is approximately .012-inch (as determined by inverting the bowl and measuring between the check ball and retainer) which allows for both quick seating of the ball and rapid refilling of the chamber.

Fuel bowls equipped with the mushroom inlet valve are externally identical to those equipped with a ball check valve. However when the accelerator pump housing and diaphragm are removed, a difference in the fuel bowl casting becomes obvious. Bowls originally intended for ball-check valving contain two perpendicular indented lines and two bosses (used to hold the ball retainer) cast into the area that serves as the upper portion of the pump chamber. When intended for a mushroom valve, the pump chamber is comparatively smooth, having only concentric valve and spring seat depressions. (Model 4010/4011 carburetors use only mushroom seals.)

Under normal operation, when the accelerator pump is actuated the pump cam raises the link lever, which pivots on a pin protruding from the side of the throttle body. This action moves the other end of the lever downward, forcing the override spring/nut and bolt assembly against the pump operating lever. The opposite end of the operating lever therefore pivots upward, compressing the spring-loaded pump diaphragm and forcing fuel out of the discharge nozzle. When the throttle is returned to idle position, spring pressure returns the diaphragm to the "at rest" position, causing the chamber to refill with fuel. (Pump cam lift and mounting location on the throttle lever determine the amount of fuel discharged per maximum pump stroke. Additional information pertaining to pump cams and pump circuit modifications may be found in Chapter 5.)

From the pump chamber, fuel travels through a passage into the metering block to the main body, and finally exits at the discharge nozzle (shooter). A check needle which seats in the main body, just below the shooter, prevents air flowing by the nozzle from drawing fuel out of the pump circuit. Models 4165 and 4175 use a slightly different arrangement, described in Chapter 5.

IDLE SYSTEM

When applied to a piece of machinery or the chronically unemployed (or sometimes to individuals who are inactively employed), the term "idle" refers to a rest or non-operating condition. However, when an internal combustion engine "idles," it is not at rest, but merely running at the slowest speed in its operating range. Some race engines idle at RPM levels that are in excess of the cruise speeds of stock engines. Idle is therefore a relative term, and "acceptable idle" is largely dependent upon the application for which an individual engine is built and tuned.

To people born on the wrong side of the generation gap, engine idle is acceptable only if it is smooth and slow, a condition that is primarily dependent upon camshaft profile. Performance enthusiasts will typically accept some amount of roughness at idle as a trade-off for increased high-RPM horsepower. Race car owners will accept a whole bunch of idle roughness. In fact they'll welcome it if it's part of the trade for more power, higher up the RPM scale. With these parameters, it is easy to understand why idle circuit calibrations differ between carburetors. It should also be obvious that modifications to the idle system will almost always be required when a carburetor is used in an application for which it was not originally intended.

Dilution of the intake charge by exhaust gases which have not been purged from the combustion chamber is one of the phenomenons of internal combustion engine operation. This condition is most severe at low engine speeds and dissipates as engine speed approaches the point of optimum operating efficiency. Low-speed dilution necessitates that idle mixtures be richer than cruise air/fuel ratios in order to maintain some degree of smoothness. Typically, as camshaft duration and overlap are increased, idle fuel jetting (idle feed restriction size) must be correspondingly increased to compensate for dilution and scavenging (where an amount of intake charge is drawn out through a late closing exhaust valve). It is for this reason that Holley double-pumpers (designed primarily for race applications) frequently contain an idle fuel calibration that is too rich for a street engine.

Given the wide variety of applications in which carburetors of equivalent performance potential

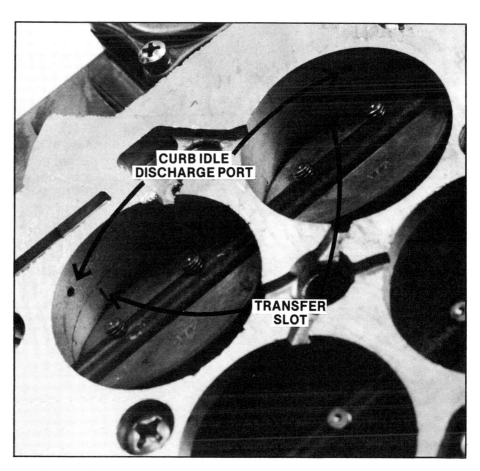

A traditional idle circuit contains a curb idle discharge port and a transfer slot. Fuel discharged through the slot is derived from the idle system but is not affected by the idle mixture screw.

are employed, it is understandable that idle calibrations can differ significantly between part numbers. Such differences also shed light on the fact that all carburetors of the same airflow capacity are not necessarily calibrated for similar applications. A 600cfm double-pumper and a "street/emissions" carburetor of the same airflow capacity will have radically different idle system calibrations.

Under normal circumstances (admittedly rare in the world of high

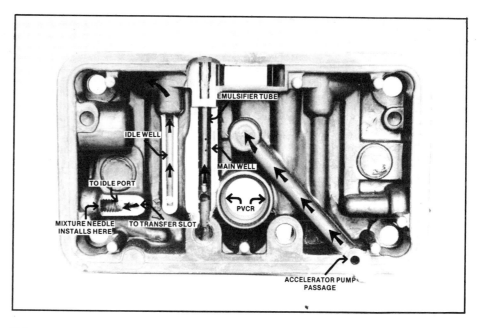

This view of the metering block shows the main well with emulsifier tube, idle well, power valve channel restrictions, and idle and transfer slot passages.

Super Tuning and Modifying Holley Carburetors

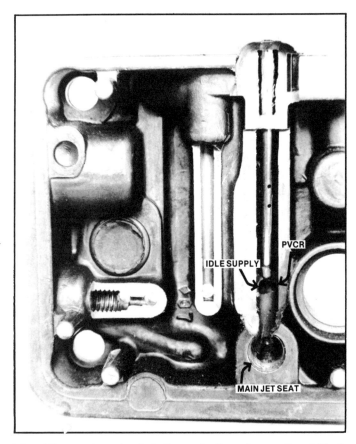

The idle system is fed through the diagonally drilled supply hole that emerges next to the idle well on the other side of the metering block. In some carburetors, the hole at the bottom of the idle well is fitted with a feed restriction.

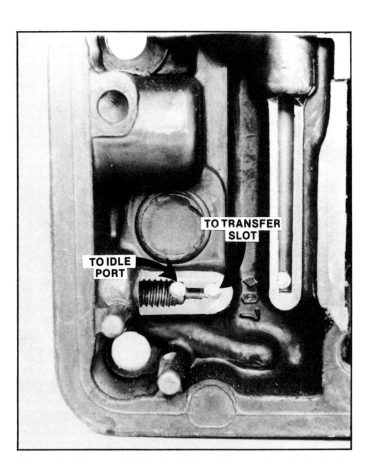

The adjusting needle for the idle mixture protrudes into the threaded hole at lower left. The passage closest to the outer edge of the metering block leads to the curb idle discharge port; the one next to it supplies fuel to the transfer slot.

A simple turn of the screw is usually all that's required to make a rough-idling engine idle smoothly. Holley uses two different types of idle circuits: One version is leaned as the screw is turned clockwise, the other is enriched. "Clockwise-lean" is referred to as standard; "clockwise-rich" pertains to late-model "reverse-idle" emissions carburetors.

performance) engine vacuum at the curb idle discharge port (located below the throttle plate) creates a fuel delivery signal much like the one that activates the main system at higher airflow rates. (In fact, the idle circuit is very similar to the main metering system and may be thought of simply as a main system for low-speed use.) Vacuum pulls fuel out of the main well through an idle feed restriction, and then through a series of passages where it is eventually emulsified with air admitted at the idle air bleed. Once emulsified, the air/fuel mixture is drawn past the mixture adjusting needle and out the idle port. The routing of idle fuel and fuel emulsion differs between carburetor models. In the Holley 4150 scheme of things, most of the routing is done in the metering block. Carburetors that are not modular in construction function along the same lines but route the idle circuit through the main body. However, in all cases idle passages typically lead fuel up toward the top of the carburetor, where bleed air is admitted, before bringing it back down to the discharge port. This must be done in order to raise the fuel above the level of fuel in the float bowl. Without the up-over-and-down routing, the discharge port would serve as a float-bowl drain.

Traditional designs locate the idle mixture adjusting screws such that the needle portion protrudes into the discharge port. Holley modular carburetors represent a departure from this practice since the mixture screws are located in the metering block, some distance from the actual port. Conceptually, the two designs are identical. The adjusting needle simply controls the amount of fuel emulsion allowed to reach the port. Placement of the needle is of little consequence.

Contrary to popular belief, the mixture needle does not control idle circuit air/fuel ratio. Although this may simply be a matter of semantics, there is a difference between the mixture in the idle circuit itself and the mixture that ultimately reaches the intake manifold. The idle circuit calibration is determined by the diameters of the idle feed restriction and idle air bleed. A "feed restriction" is nothing more than a metering jet for the idle system and the air bleed serves as an airflow regulating orifice. As the air/fuel emulsion exits the idle port it is mixed with small amounts of intake air that flows past the throttle plates. Varying the throttle plate opening (with the idle speed screw) increases or decreases both the amount of air that is allowed to flow into the intake manifold and the amount of pre-mixed air/fuel emulsion discharged at the idle port. Since these two idle mixture components are affected by degree of throttle opening, the ratio of intake air to idle mixture air/fuel is not drastically altered by changes in idle speed. Turning the idle mixture screw varies the volume of air/fuel emulsion discharged into the manifold, not the ratio of air to fuel in the emulsion. Therefore, when a satisfactory idle cannot be achieved by "tweaking" the mixture screws richer (more idle air/fuel to intake air) or leaner (less), it can be assumed that the carburetor idle calibration is not within the range demanded by that particular engine.

Use of two four-barrel carburetors is an excellent example of such incompatibility. With such an installation, the volume of air/fuel emulsion is effectively double that available from a single four-barrel mounted on the same engine. Even when the mixture screws are adjusted toward lean, there may simply be too much emulsified idle fuel flowing to allow a clean idle. In such cases, the idle feed restrictions of both carburetors must be reduced, in cross section size, to lean the idle mixture to an acceptable level.

Idle mixture calibrations are rather crucial, especially in these days of stringent emissions regulations. Since the original settings rarely need alterations for standard applications, most manufacturers supply carburetors with fixed metering restrictions.

When idle system calibration changes are required, the restrictions must either be drilled larger or removed and replaced with ones containing a smaller diameter orifice. These types of alterations are confined almost exclusively to highly specialized race applications. Satisfactory idle performance can, for the most part, be achieved by varying the settings of the idle mixture screws, and the idle speed screw.

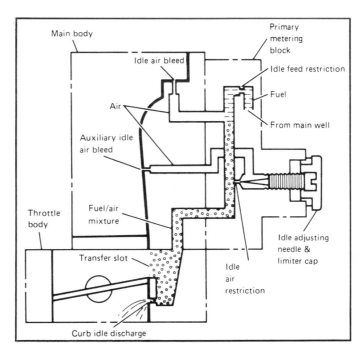

The reverse-idle system is actually a variable air-bleed arrangement. Idle fuel mixture is controlled (richer or leaner) by variations in the amount of bleed air admitted to emulsify the idle fuel.

On reverse-idle carburetors, such as the 6979, idle fuel emulsion is delivered to the idle discharge and the transfer slot through the same passage. The separate passage used for the transfer slot in standard-idle carbs is not used. The idle-adjusting screw controls emulsion delivery to both the idle discharge and transfer slot.

REVERSE IDLE SYSTEMS

Typically, rotating an idle mixture screw counterclockwise (on just about any carburetor) means that more air/fuel mix would be discharged and a richer idle mixture would result. In the late 1970s, carburetor manufacturers found that a more sophisticated idle system had to be devised in order to meet emissions standards. Holley's approach was to incorporate a second adjustable air bleed into the idle system. By changing a few passages in the main body, the mixture screw in the metering block was used to adjust bleed air, rather than fuel. Carburetors incorporating this system are commonly referred to as "reverse-idle" models, since the engine receives a richer mixture when the screws are turned clockwise and a leaner mixture when they're rotated counterclockwise. A "reverse-idle" carburetor is identified by a silver tag containing an arrow and the word "lean" above the mixture screw.

Implementation of the "reverse" idle system is quite simple. In a conventional carburetor two separate

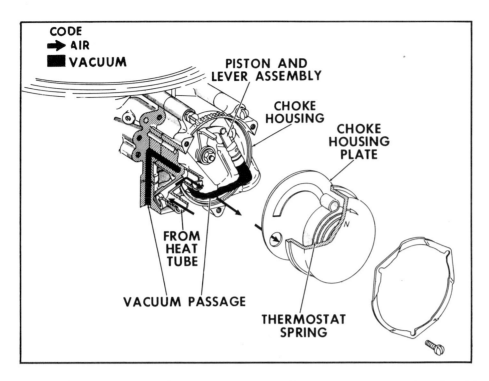

Integral (and some electric) chokes use a manifold vacuum to circulate heat (or air) through the choke housing or to activate the pull-off piston. Incorrect mounting of choke system components therefore can lead to serious vacuum leaks.

passages are used to supply air/fuel mix to the discharge port and the transfer slot. When viewing a Holley main body's metering-block mating surface, two sets of two-cored holes are visible on either side of surface. The outer holes lead to the idle port of each throttle bore and the inner holes lead to the transfer slots. Traveling in the other direction (toward the float bowl), the holes in the main body mate to fuel supply passages in the metering block.

As fuel flows up out of the idle well and across the connecting passage into the idle passage, it eventually reaches the main body hole that leads to the transfer slot. A hole at the bottom of the down passage (in the metering block) connects to a short passage that leads to the needle end of the mixture adjusting screw and yet another hole. This second hole connects to the outer hole in the main body and supplies fuel to the discharge port.

Normally, as the mixture screw is rotated counterclockwise, the small passage in the metering block is opened wider and more air/fuel mix is channeled to the idle discharge port. Turning the screw clockwise has the reverse effect and when the needle is firmly seated against its seat, no fuel reaches the idle port.

On the "reverse-idle" carburetor everything is the same except that, rather than leading to the idle port, the cored hole in the main body leads to an auxiliary air bleed in the base of the venturi. The idle discharge port is therefore supplied by the same channel that leads to the transfer slot. Fuel flow to the port is not adjustable; the mixture screw is now used to control the amount of secondary bleed air admitted to the idle system. This is why it is impossible to kill the engine when adjusting the idle mixture on a "reverse idle" carburetor. Regardless of the screw setting, the engine will be supplied with some amount of idle fuel.

At 12:1 to 14:1 the ratio of air to fuel is rather lopsided. Not that it shouldn't be that way, considering the cost of each, but the disparity calls for some amount of consideration when circuit alterations are conceived. In the case of the reverse-idle system, the tip of the needle had to be recontoured so that more adjustment was obtainable per turn of the adjusting screw. Whereas the end of a standard idle mixture screw is machined into a long tapered needle, the "reverse-idle" screw is more like a blunt wedge. Were the standard needle

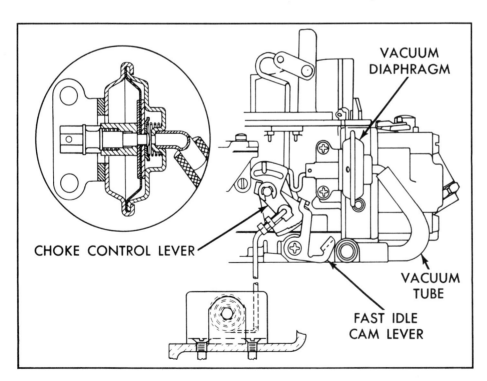

Divorced chokes use a bi-metal spring attached to the manifold to pull the choke plate completely closed and pull the fast-idle cam into place. As soon as the engine starts, the manifold vacuum applied to the vacuum diaphragm pulls the choke plate open slightly (called the choke "qualifier") to allow enough airflow for satisfactory idle.

configuration used to control bleed air, the screw would have to be rotated one or more turns in order to affect any appreciable change in mixture.

Just as the term "economizer valve" is a misnomer, "idle system" is somewhat misleading. The circuit does function at idle, but it also serves to bring an engine off idle and into a higher speed range. (Not wishing to commit another nomenclature *faux pas*, I won't attempt to rename the circuit. Just remember, you read it here first.) As the throttle plates are opened, the portion of the transfer slot exposed to manifold vacuum increases, bringing about a corresponding increase in fuel flow. Eventually, the throttle plates will be opened sufficiently to actuate the main system and flow through the idle circuit will diminish. The purpose of the transfer slot is, as its name implies, to transfer fuel control from the idle to the main system.

Although there is no provision to "shut-off" idle fuel once the main system is activated, this is effectively what occurs. Once the throttle plates are open far enough to allow passage of a significant quantity of air, the amount of vacuum presented to the idle port and transfer slot has been reduced. Concurrently, a strong signal has been developed at the main discharge nozzle, so fuel that might otherwise reach the idle system is "stolen" by the main circuit. (Both idle and main systems are supplied by the main well.) Since idle fuel is no longer needed, flow effectively ceases.

CHOKES AND COLD STARTING

When the cold of night (or day, for that matter) strikes, it isn't just we mortals who suffer discomfort. The internal combustion engine is right beside us, shaking and shuddering, albeit for different reasons. It is a simple fact of automotive life that gasoline is not readily atomized when it and the metallic passageways through which it must flow are cold. Add to these considerations the comparatively low speed (and attendant low vacuum) at which a starter spins an engine and it can be seen that cold starting does not provide ideal conditions for engine operation. Without a good strong signal, very little fuel is drawn out of the carburetor, and most of that is returned to a liquid state as on contact with a cold manifold runner.

A choke plate is the device used to alleviate problems associated with coaxing a cold engine to life. When the plate is closed, a strong vacuum is created immediately below it (even by a slow-turning engine) and fuel is drawn out of the idle and main systems. With very little air and an abundance of fuel, conditions are highly conducive to combustion and an engine will usually spring to life. Once this occurs, there is sufficient vacuum to initiate fuel flow and cause a reasonable amount of atomization. Fuel will still condense on the intake manifold walls, but a richer-than-normal mixture is continually supplied until the manifold warms up and the choke is fully open.

The choke plate functions in conjunction with a fast idle cam that raises engine speed when the choke operates. Most automatic choke systems are also fitted with a "pull-off" diaphragm which partially opens it when vacuum develops at engine start to prevent an overly rich idle.

Several types of choke systems have been used over the years. Older carburetors were equipped with an integral choke, wherein hot air (sometimes from the exhaust manifold) was piped directly to the bimetal thermostatic spring. Vacuum was also applied to the choke housing so that the hot air was drawn into the spring chamber. This type of choke can be a problem during hot starts because the thermostatic spring will cool considerably faster than the engine, allowing the choke to close even though the engine is still warm enough for a choke-open start. Deterioration with age is another problem. Carbon build-up in the hot air tube can reduce or completely stop the flow of warm air to the spring, resulting in a choke that either opens very slowly or not at all.

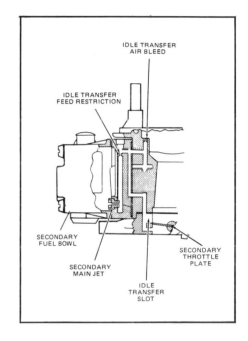

A secondary idle system is used on Model 4150 four-barrels to prevent fuel stagnation in the secondary float bowl. Emulsified fuel flows from the idle passage below the transfer port at curb idle. As the secondary throttles open, fuel also begins to flow from the transfer port.

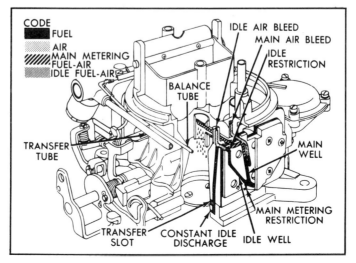

The secondary main metering system of Model 4160 functions in the same manner as Model 4150. A metering plate is used rather than the more complex metering block. Passages are visible when the plate is removed.

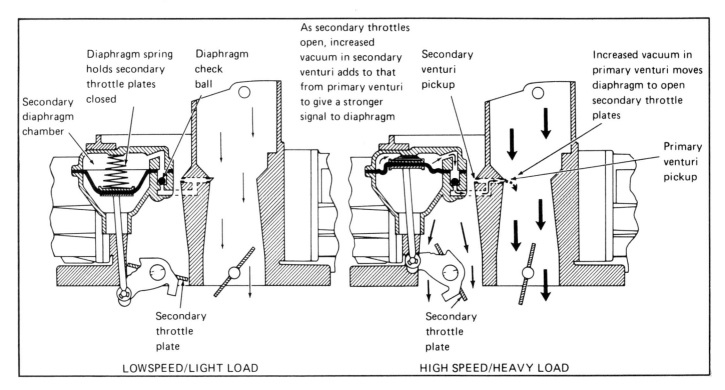

An alternative to the integral choke is the divorced type, wherein the thermostatic spring is mounted either in a well or on a pad on the intake manifold, just above the exhaust crossover passage. A rod connects the bimetal spring to the choke control lever, and a "pull-off" diaphragm is used much in the manner that it is employed on an integral choke.

A primary difference between a divorced and integral type of choke is that the former lacks adjustability. The phenolic housing used with the integral choke is typically marked "rich" and "lean." Rotating the housing in the "rich" direction increases the spring preload, causing the choke plate to open more slowly. Turning the housing toward "lean" logically has the opposite effect. Some early divorced choke bimetal units had a provision for similar adjustment but this was largely abandoned on later models.

Use of electric chokes is confined almost exclusively to aftermarket replacement carburetors which must fit a variety of intake manifolds. An electric choke is an integral type except that the spring housing contains a heating element that is connected to a 12-volt ignition switch-controlled power source. When the car is started and electri-

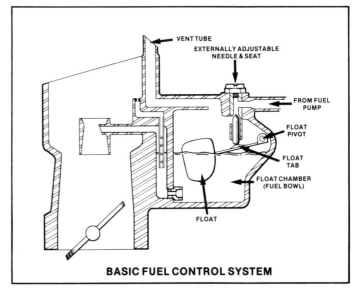

BASIC FUEL CONTROL SYSTEM

Vacuum-operated secondary throttles also are controlled by the Venturi Effect. As in-rushing air passes through the venturi (high-speed/high-load conditions), a pressure drop is created. This drop (vacuum) creates low pressure above the secondary vacuum diaphragm, which is then forced upward by the atmospheric pressure below the diaphragm.

cal current is applied, electrically induced heat causes the same type of bimetal reaction created by exhaust heat. The electric choke allows for a great deal of versatility on custom carburetor/intake manifold installations. However, the coil or other ignition components should not be used as an electrical power source since the current drain would adversely effect the system operation.

Prior to the days of "high tech" carburetor development, a hand choke was the only alternative to extremely difficult cold starting. The beauty of a hand choke is its simplicity; a cable with a knob attached to one end controls operation. It is the person behind the wheel who controls the choke completely and he or she can decide whether a quick or slow opening is appropriate. The driver can also forget to open the choke as the engine warms. This inattention will result in black smoke belching from the tailpipe and in an immediate drop in fuel economy.

In warmer weather, it's possible to drive without a choke of any type. Cold starting is accomplished by depressing the gas pedal several times so that the accelerator pump squirts a healthy volume of raw fuel into the intake manifold. This enables the engine to start, but it will run very erratically until it warms enough to allow normal atomization.

SECONDARY METERING SYSTEMS

In all Holley four-barrel carburetors, the secondary metering circuits are almost mirror images of the primaries. The differences lie not in concept but in manner of execution, as demanded by operational parameters. The secondary barrels are used only a fraction of the time (except by overly enthusiastic types), therefore special features must be included to accommodate comparatively rare and intermittent use.

SECONDARY IDLE CIRCUIT

A prime example is the secondary idle system. A sufficient amount of idle fuel is passed through the primary circuit to maintain engine operation, so it would appear that a second idle circuit is unnecessary. However, if no fuel was taken from the secondary side at idle and if a driver was very conservative in the application of the accelerator pedal, fuel in the secondary bowl would be allowed to turn stale. After a prolonged period of non-use, varnish and gum would begin to block various passages and when the secondaries were finally called into action, fuel delivery would be inadequate. By including a secondary idle circuit, fuel continuously flows through the float bowl and out the constant-feed discharge hole (the secondary equivalent of the curb idle discharge port). This insures that the secondary fuel bowl is always filled with fresh fuel and that the inlet needle valve does not become "varnished" closed. Secondary idle fuel also helps improve fuel distribution at low engine speeds.

Since the secondary idle circuit is not essential to engine operation it usually has no provision for mixture adjustment. An exception to this may be found on certain race carburetors where adjustment provision is included to ease tuning of engines with long-duration camshafts. Except for the fact that this system passes less fuel and is not (generally) adjustable, it is identical to the primary version.

SECONDARY MAIN METERING CIRCUIT

With the exception of Models 4160 and 4175, which use a metering plate rather than a block with removable jets, the primary and secondary main metering circuits are identical. Details concerning fuel flow through the secondary metering plate may be found in the drawing on page 36.

POWER ENRICHMENT

Since the secondary throttles are used only under full power and high-load conditions, many carburetors contain relatively rich secondary jetting and are not fitted with power valves. When secondary power valves are included it is usually the result of an effort to reduce jet size (in order to minimize fuel slosh and spill-over on deceleration). Some race carburetors are fitted with secondary power valves when fuel flow through the largest available jets is insufficient. The need for supplementary fuel is usually confined to large displacement engines fitted with a single Model 4500 carb.

On Model 4160 and 4175 carburetors, there is no provision for a secondary power valve since a metering plate rather than a block is used. The main metering restrictions in the plate are drilled on a vertical plane, rather than horizontally, minimizing problems with fuel slosh through the restriction on deceleration.

SECONDARY ACTUATION

VACUUM SECONDARIES

If the secondary throttle plates were linked directly to the accelerator pedal, oil-company profits would be greater than they are now. Constant use of all four barrels is not only wasteful, but would also keep most engines in a state of over-carburetion. When Holley engineers designed the 4150 series of four-barrels, they elected to use a vacuum diaphragm, which reacts to airflow, to operate the secondary throttles. Other manufacturers (Carter, Rochester) use an air valve to tie secondary opening to engine airflow requirements. Irrespective of the method employed, the concept of vacuum activated secondaries is to provide only enough airflow to fulfill engine requirements. Some form of spring is normally used to control the opening rate of the secondaries. In a Holley, this spring works in conjunction with the opening diaphragm. In a Rochester or Carter, a spring or counterweight is used to preload the secondary air valve.

On a Holley, when airflow through the primaries reaches a given point, vacuum created in a pickup tube in the primary venturi is applied to the top portion of the diaphragm, pulling the secondary throttle plates open. As engine speed (and consequently airflow) increases, so does vacuum at the venturi, causing further raising of the diaphragm. And once a reasonable amount of air starts flowing through the secondaries, additional vacuum (applied through a pickup tube in the secondary venturi) is exerted upon the diaphragm, insuring that it fully opens the secondary throttle plates. At low engine speeds, the secondary pickup also functions as a bleed to prevent secondary actuation when not required. The bleed, occasionally in combination with a check ball in the diaphragm housing, allows smooth, steady secondary opening. Upon closure of the primary throttle plates, the vacuum applied to the diaphragm drops and, with no force to overcome spring pressure, the diaphragm is pushed down, rotating the secondary throttle shaft to a closed position.

Considering that the primary purpose of the secondary diaphragm spring is to control opening, not closing, of the throttle plate, the

Super Tuning and Modifying Holley Carburetors

foregoing explanation may seem like somewhat of a back-door approach. It is, but a detailed explanation of the relationship of the spring to opening rates is contained in the next chapter.

MECHANICAL SECONDARIES

As the name implies, "mechanical secondaries" operate through a direct (or mechanical) link to the accelerator pedal linkage. On occasion, mechanical actuation is referred to as progressive. This description bears no relevance to operation but refers to the interrelationship of primary-to-secondary opening points and rates. Typically, the primary throttle plates will be open 40-45 degrees before secondary actuation is started. This allows for good low-speed driveability and fuel economy. However, due to the fact that the primaries are already open 40 degrees or more before throttle opening progresses to the secondaries, the opening rate of the latter must be much quicker so that the primaries and secondaries reach wide-open-throttle at the same time. Direct or 1:1 linkage is used only on a few "race-only" carburetors such as the 0-6214, 0-6464, and 0-4224.

Under most operating conditions, rapidly rotating the secondary throttle plates from closed to fully open would result in a severe stumble if an additional quantity of fuel was not also immediately made available. This is the same situation that exists on the primary side, and it is solved the same way — with an accelerator pump. The two circuits are virtually identical with the exception being certain part numbers that use the larger 50cc pump. Only Holley's Model 4360 is equipped with mechanical secondaries and no secondary accelerator pump.

FUEL INLET CONTROL

Back in the days of the Model T, which is, thankfully, long before my time, the accomplished motorist carried a pair of pliers, some bailing wire, and a screwdriver. So equipped, he or she could fix all but the most serious engine problem. Most garages tackled serious repairs using only slightly more sophisticated tools such as hammers. And it was not uncommon to see a "tuneup expert" using the business end of a hammer against a float bowl, as the hammerer attempted to free an inlet valve that had become stuck. Thus was born the forerunner of feedback carburetion.

As fuel quality, carburetor manufacturing techniques, and metallurgy improved, the need for hammers and the like diminished. The float/inlet valve configurations currently employed by carburetor manufacturers typically provide trouble-free operation for thousands of miles. The inlet valve (commonly called the needle-and-seat) size and shape differs among carburetor models. However, operational characteristics are virtually identical whether the carburetor in question is a Holley, Carter, or Rochester. Either a fuel pump, or gravity, causes fuel to flow from the gas tank to the carburetor inlet. Many carbs have a smaller filter or filtering screen at the inlet to remove dirt and other particles. Once through the filter, the fuel flows past the inlet valve into the float bowl. As the volume of fuel in the bowl increases, the float rises and with it goes the inlet needle (occasionally it is clipped to the tab that extends from the float body to the pivot pin). Ultimately, the float is lifted high enough to force the needle firmly against the seat, thereby stopping the flow of fuel into the chamber.

Fuel exits the float bowl through the main jets and when enough fuel has taken leave, the float drops, opening the valve and allowing the level to rise until the needle is once again forced firmly against the seat. Adjustment of fuel level is accomplished by altering the position at which the float causes valve closing. On Holley 4010/4011 and 4150/4160/4500 carburetors, fuel level is externally adjustable. By loosening the lock screw and rotating the adjusting nut, the needle-and-seat assembly may be raised or lowered, with a corresponding change in fuel level. Most other carburetors use an internally adjustable float. Alteration of fuel level is accomplished by bending the tab attached to the pivot pin. Bending the tab so that the float is moved downward lowers fuel level; raising the float in relation to the pivot has the opposite effect.

(In some carburetors, float drop is also adjustable. For high-performance use, drop is usually set beyond specifications to insure an adequate flow of fuel past the needle. Insufficient float drop will place the needle in the path of incoming fuel, creating a flow restriction even when the float bowl is empty.)

Proper adjustment of fuel level is critical to satisfactory engine operation. If the level is too low, activation of the main system is delayed; fuel mixtures can become too lean and the engine may temporarily run out of gas. Too high a level can lead to over-richness, flooding, and fuel slosh out the vent tube.

FLOAT BOWLS

In the 4150 scheme of things, float bowls, floats and needle-and-seats vary according to the original application intent of a carburetor. The "flat" or side inlet float bowls are used for most medium performance and emissions-era street carburetors. These bowls are usually fitted with a transfer tube that delivers fuel from the primary side inlet back to the secondary. However, on some race applications, where use of the side inlet bowls is necessitated by clearance problems (inline, dual four-barrels), provision is made for each bowl to be fed by a separate fuel line.

By far the most popular float bowl is the center inlet or "dual-feed" model. Placement of the float pivot in the center of the bowl, rather than off to one side, alters the relationship of side-to-side fuel movement and float bowl fuel volume. With the side inlet bowl, whenever the vehicle deviates from straight line travel and centrifugal force pushes most of the fuel in the float bowl to one side, the direction in which the vehicle is turning has a significant effect on fuel level. Concentrating the fuel close to the pivot pin allows a greater volume of liquid to enter the float bowl because leverage is reduced and inlet valve closure is delayed. Conversely, when fuel moves away from the pivot point, leverage is increased, causing premature closing of the inlet valve and a reduction in fuel volume.

The advantage of the center inlet bowl is that side-to-side fuel movement does not have so dramatic an effect on float bowl volume. Moving the fuel to either side of the float chamber has the same effect, since either extreme is equidistant from the pivot point. However, the "dual-feed" bowl is more sensitive to fore and-aft fuel movement because this is where the leverage deviations occur.

In years past, Holley offered a few unique float bowl designs for certain high-performance and race carburetors. The currently available side- and-center inlet bowls have proven most satisfactory for virtually any application, hence the more exotic configurations are no longer produced. All Model 4010/4011 carburetors are of the dual-feed design with center-pivot floats.

SETTING OF FUEL LEVEL

Fuel level may be checked and adjusted through a variety of procedures, as determined by carburetor type. Where internal adjustment is employed, a measurement from a reference surface is commonly specified by the manufacturer. This specification is checked by removing the float bowl and measuring the relative distance between the top of the float and a reference surface in the bowl (see Chapter 5).

With an externally adjustable float, disassembly is not required, and level can be checked with the engine running. This manner of adjustment is more accurate since fuel level may be verified under operating conditions. It is also much easier. With the sight plug in the side of the float bowl removed, the adjustment nut is turned until fuel just reaches the bottom threads of the sight plug hole. Engine RPM can also be increased to determine if fuel pressure variations are affecting the level.

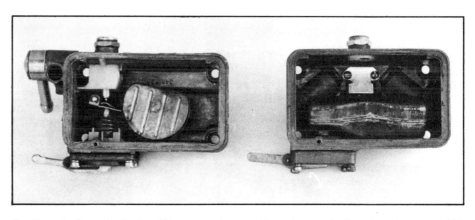

Fuel control, or the lack of it, can make an otherwise perfect carburetor seem like a candidate for the scrap heap. Both the side inlet and center inlet fuel bowls offer external float adjustment. A center inlet design is preferred for applications where high-speed cornering is common, such as when driving enthusiastically to the corner market.

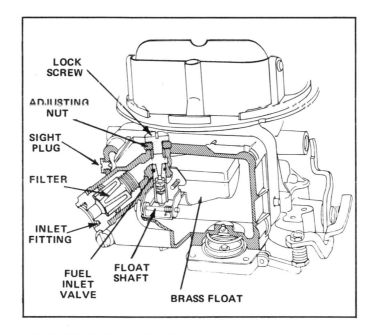

One of the main advantages offered by the Holley modular design is an externally adjustable fuel level.

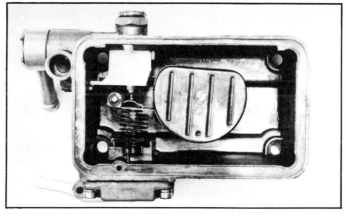

The spring below the float tab is designed to dampen float oscillations and to ensure positive closure of the inlet needle when the fuel level rises to the proper level.

Super Tuning and Modifying Holley Carburetors

Super Tuning and Modifying HOLLEY CARBURETORS
Basic Modification Concepts

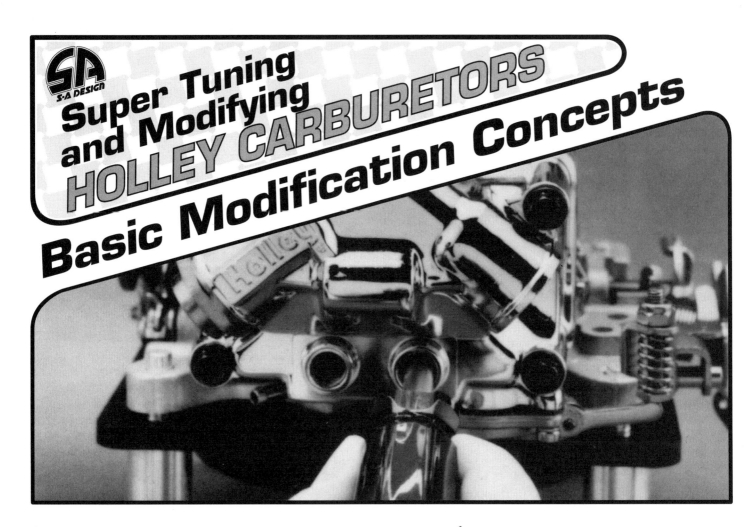

There is virtually no limitation to the degree and type of carburetor modifications performed by racers and performance enthusiasts. There is also virtually no limitation to the degree of effectiveness achieved by these modifications. Some "tuneup tips" work just fine when all of the throttle blades are standing on end (wide-open throttle) and some don't. The key to carb tuning success is knowledge — knowing what works and what doesn't. Knowing how to modify a carb is not enough; understanding the "whys" and "hows" that accompany the "wherefores" is essential so the effectiveness of a modification can be accurately evaluated.

For openers, *the original application for which a carburetor is intended dictates the parameters for all fuel delivery calibrations.* Since each carburetor delivery circuit interacts with others, a seemingly simple modification may have numerous ramifications. The end result of such modifications may be a carburetor that suffers from "terminal" disfunction. Subsequent chapters deal with modifications of specific carburetors for specific applications. However, the number of possible "custom" carburetor installations is almost infinite, so this chapter is devoted to explanations of a few Holley design philosophies. It should be considered essential reading. Though you're undoubtedly more interested in the hands-on modification information, the information on the following pages will be extremely useful. Besides, since I took the trouble to write it, you should be courteous enough to read it.

PRIMARY IDLE SYSTEM—STANDARD

The extensive interchangeability and similarity of components makes it rather difficult to categorize 4150/60 series carburetors. Nonetheless, there are two types of idle systems that exist in the 4150/60 series, and they differ considerably. The standard system employs a "clockwise-lean and counterclockwise-rich" idle mixture adjustment and is found on most race and "non-emissions" high-performance models. The second type is called the reverse-idle system and is found primarily on emissions-type models (more about this system later).

Adjustment of idle mixture using the standard "clockwise-lean" adjustment, and the resulting effect on engine performance, is fairly straightforward. As the adjustment needle is turned clockwise, it moves further into the passage at the bottom of the idle down-well that connects the idle discharge port and the idle transfer passage (see illustration in Chapter 4). The connecting passage functions as a seat. When the needle is turned as far as it will go in a clockwise direction, the flow of air/fuel emulsion to the curb-idle discharge port is stopped. In effect, air/fuel emulsion flowing in the idle down-well follows a "Y" routing, proceeding to either the curb idle discharge or the idle transfer slot portions of the circuit. The mixture needle is on the curb-idle leg of the "Y" and the idle-transfer passage is on the other. Consequently, adjustment of the idle mixture affects gas mileage only to the degree that it controls the amount of fuel delivered by the curb-idle discharge hole. (This is in marked contrast to the effects of needle adjustment on a reverse-idle circuit.)

In cases where the main system has been leaned as much as possible, yet there is still an excessively rich mixture at low engine speeds, it will be necessary to reduce the size of the idle feed restriction or enlarge the diameter of the idle air bleed. Bleed size is based on several things: The diameter of the curb-idle port; the desired position of throttle plates at idle; and the idle-transfer circuitry design. Alterations of the bleed should be a last resort, since such changes affect system timing.

Where possible, the idle feed restrictions should be altered whenever it's necessary to recalibrate the idle circuit. In some metering blocks, the restrictions are located within the idle wells and replacing them is difficult or impossible without the proper tools and replacement well-plugs. However, in many blocks the restrictions are located at the base of the idle well and are accessible from the main-body side of the block. In these cases, the brass restriction may be replaced with a smaller restriction, or a V-shaped wire can be inserted (see photo) to restrict flow, thereby leaning the idle mixture.

As with any metering orifice, the total area of the orifice, rather than its diameter, is the critical factor. And, since area is a function of diameter, it should be noted that when dealing with a restriction as small as .030- or .035-inch in diameter, changes of .001-inch are significant. Increasing diameter from .030- to .031-inch represents an area (and flow) increase of almost 7%. All changes in area should be computed with the equation, Area = Pi (r2). Examples of such computations are contained in Chapter 6.

In most street applications, modification of the idle circuit is usually necessary only when a double-pumper or other race-type carburetor is used, or when a radical camshaft is employed and manifold vacuum at idle is extremely low. Race carburetors are typically fitted with a rich idle calibration as a means of compensating for exhaust dilution (which is a function of cam timing). When such

MODIFICATION	EFFECT
Jet Change	Increase jet size (higher number) = richer. Decrease = leaner mixture
Air bleed diameter	Enlarging bleed leans mixture and delays activation of fuel discharge at the nozzle. Reducing bleed size enriches mixture and starts main system fuel flow at lower air flows.
Booster Venturi Replacement (requires special booster installation tool) 4150/4160 Only	Booster configuration has a direct effect on pressure drop in the Venturi. The most restrictive booster produces the highest fuel delivery signal. The reverse is also true. Down-Leg - least restrictive Straight Truck #1 Truck #2 Economaster - most restrictive
Throttle Body Replacement (requires blending of bores in main body to match new throttle bore diameter).	Increasing throttle bore size without altering venturi size or booster configuration increases pressure drop and therefore signal strength. The reverse is also true.

It is relatively easy to modify Holley two- and four-barrel carbs, but knowing how and why a specific modification is performed is the real key to success.

On race cars such as this Pro Stocker, extensive modification of carburetors is standard practice. However, the worth of such rework has never been quantified, since most racers have never run "stock" carburetors. That isn't meant to imply that stock carburetors should be run on race cars; it's merely intended to point out that, in many instances, the performance improvement derived from carb reworking is not known. Therefore, some claims may be overstated.

On 4150/4160 Holley carburetors the idle-mixture screws are located on either side of the primary metering block. On standard idle carbs the screws are turned clockwise to lean the mixture and counterclockwise to enrich the mixture.

About 100 years ago, when he ran Super Stock, Bill Jenkins laid down impressive numbers with only minimal changes to the carburetor.

On some emissions-type Holley carbs, a reverse-idle adjustment system is used to gain increased sensitivity to idle mixture control. With the reverse-idle system, the screw regulates bleed air flow rather than fuel flow.

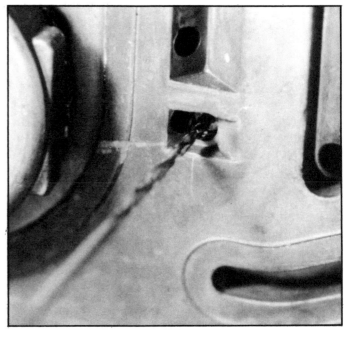

Alteration of the idle feed restriction is only necessary when a carburetor is installed in an application for which it is not designed. If a street carb is installed in a radical race engine, the idle feed restrictions may have to be enlarged to increase idle fuel flow.

carbs are used on a mildly cammed engine, idle mixtures may be far richer than necessary. (In fact, such carburetors are occasionally too rich for a race engine.)

Conversely, a street-calibrated carburetor may have insufficient idle fuel delivery for an engine equipped with a long-duration camshaft. In the latter case, increasing the diameter of the idle feed will improve both idle quality and off-idle performance. This approach can be very effective in eliminating the off-idle stumble that is characteristic of an engine equipped with a large-runner, single-plane manifold and a long-duration camshaft.

It should also be noted that the air/fuel ratio of the mixture delivered through the transfer slot may play a major role in determining off-idle performance characteristics of street-driven vehicles. If a car is geared such that the throttle blade angle is very low at normal cruise speeds, fuel delivery will be primarily through the idle transfer slots/ports because the throttle plates aren't open sufficiently to initiate significant air flow through the booster venturi/main circuit. As a result, fuel economy will take a serious turn for the worse if the idle feed restriction is extremely large.

PRIMARY IDLE SYSTEM—REVERSE

This system is another testimony to the versatility of the Holley modular design and the craftiness of Holley engineers. By merely drilling a few extra holes and eliminating some others, a vastly different type of idle circuit was developed. In reverse-idle carburetors, the traditional curb idle discharge portion of the circuit has been eliminated. All idle emulsion is delivered through the routing that (in a standard system) normally supplies only the transfer slot. (A small constant-feed hole drilled into the pocket that supplies the transfer slot replaces the curb idle hole.) The passage in the main body that formerly fed fuel to the curb idle port now leads to an auxiliary bleed hole in the base of the venturi. Turning the idle-mixture screw fully clockwise completely closes the path to the auxiliary bleed and brings the idle fuel mixture to maximum richness. Turning the screw in the reverse direction

The primary idle air bleeds (the larger bleed shown here) and the high-speed air bleeds are located inside the choke shroud. If a suitably lean idle mixture cannot be obtained by adjusting the idle mixture screws, as a last resort the idle air bleeds can be enlarged slightly to lean the idle mixture.

Some carbs offer improved performance when the idle feed restrictions are made smaller, and consequently the idle mixture is leaner. A V-shaped wire serves nicely to reduce fuel flow. This method also can be used to save metering blocks that have had the restrictions drilled too large by an overly enthusiastic modifier.

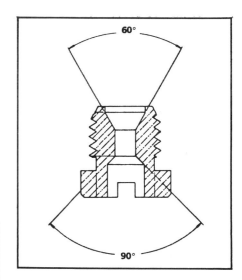

The typical Holley jet is chamfered on both inlet and outlet sides. Drilling a standard jet oversize removes portions of this chamfer, thereby lengthening the metering passage and increasing the amount of flow restriction. When this occurs, a larger-diameter drilled jet actually may flow less than a smaller standard jet.

On Holley 4150 family, 4500, and 2300 carbs, the main metering system easily is tuned to specific requirements. The main jets (arrows) are accessible by removing the primary float bowl. Changing the main jets primarily will affect overall fuel-air ratio at virtually all engine speeds and loads.

(counter-clockwise) admits more bleed air for a progressively leaner idle mix. The reverse-idle system is employed only in "emissions-type" carburetors; the air-bleed idle system was developed specifically to improve idle quality and keep exhaust emissions within predetermined levels.

In the reverse-idle system, modification of the idle feed restriction constitutes the only practical means of altering idle calibration beyond the existing limits. Drilling the idle-air bleed is senseless because air bleed adjustment is already provided by the idle-mixture screws (which are now bleed-air screws). However, before altering the idle feed restrictions in a street-driven application, realize that any enriching of the idle mixture can seriously alter the overall fuel economy (of a reverse-idle carb). With reverse idle circuitry, idle-emulsion delivery through the transfer slot is influenced by the idle feed restriction **and** by the mixture-adjusting screws. (In a standard-idle circuit, the adjusting screw only affects curb idle air/fuel emulsion delivery.) And since the transfer slot delivers emulsion during early off-idle/acceleration/cruise operation, particular care should be taken to insure that the idle emulsion is adjusted as lean as practicable to maximize fuel economy and minimize exhaust emissions.

When other considerations don't preclude such a swap, it is possible to interchange standard and reverse-idle metering blocks. However, adjustability will be adversely affected. Since roughly 13 times as much air as fuel is consumed in an idle mixture, the air bleed adjustment must be much coarser than the mixture adjustment in a standard carburetor. For this reason, reverse-idle metering blocks contain adjusting screws with comparatively blunt tips, whereas a conventional idle needle is tapered to a fine point. These adjusting screws are physically interchangeable, but quality of adjustment will be less than optimal.

Installation of reverse-idle carburetors on high-performance engines can pose a problem. Without reworking the idle circuit, fuel delivery is sometimes inadequate to compensate for exhaust dilution of the intake charge. Reverse-idle circuitry is used on some models of both the 4150/4160 and 4165/4175 families. The 4010/4011 series of carburetors employs a standard "clockwise-lean" idle circuit.

MAIN METERING SYSTEM

In theory, if the right carburetor is chosen for a particular application, no changes in the main metering circuit, other than jets, should be necessary. However, in some highly specialized race environments, it may be of value to change booster venturis, throttle bodies, or air bleed diameter in an effort to increase air flow and tuning capability. Unfortunately, there is little

Super Tuning and Modifying Holley Carburetors **43**

magic to be had from an overactive imagination applied through oversized drill bits. Without flow equipment to determine exact fuel delivery curves, any main metering system modification is usually the equivalent of a shot-in-the-dark. This isn't to say that improvements are impossible. Many of the people now considered carburetor modification "experts" stumbled around in the dark for many years before tripping over a combination of the right holes drilled in the right locations. Some of these experts are still stumbling.

In any event, a systematic approach and accurate record-keeping are essential so successful modifications can be duplicated and failures can be hidden in a closet. Considering that improper main metering system modifications can ruin a perfectly good carburetor, (unless the modifier has access to such items as air bleed restrictions, a booster installation tool and a variety of hard-to-get small parts) a conservative approach is the best means to minimize the creation of "closet queen" carburetors.

The most common "trick" modifications and the resultant effects are listed in this chapter on the accompanying Modification/Effect chart. Note that the effects on fuel delivery are quite difficult to detect. If you ever find yourself considering any of these modifications, the odds are overwhelming that you are trying to force the carburetor to work in an entirely wrong application. Save your time and money — get the right carburetor and eliminate the need for Band-Aid modifications.

In addition to the alterations listed in the chart, the only practical main system modification consists of enlarging the channels and wells in the main body and metering block to accommodate extremely high fuel flow requirements. When using gasoline, such reworking is generally not required for ordinary usage. Model 4150/4160 carburetors need to have fuel passage modification only when used with alcohol. (Modifications for carburetors used on alcohol-burning engines are covered in a subsequent portion of this chapter.)

Changing the main jets is the most commonly employed form of carburetor recalibration. It is a good practice to purchase an assortment of jets, rather than drilling an existing jet to a larger size. Holley rates its jets by flow capacity rather than diameter. By adjusting entry and exit angles, a jet of a given drill size can be given a variety of flow characteristics. As an example, number 88, 89, and 90 jets all contain a .104-inch diameter hole. Close examination of a variety of Holley jets reveals that the chamfer leading to the orifice differs with jet number. What may not be quite as obvious is that the length of the orifice also varies. As this length is increased, a jet of a given diameter becomes more restrictive.

It is entirely possible that a jet redrilled to a larger diameter will in fact lean the mixture rather than provide the desired enrichment. If the front and/or rear side of the jet is chamfered, enlarging the orifice will remove all or a portion of the chamfers, thereby increasing the length of the restriction and reducing flow capacity. Additionally, removing the entry chamfer exposes a sharp edge which creates turbulence as the fuel attempts to flow into the metering orifice. The jet chamfer performs the same function as a velocity stack, that of smoothing the entry area to reduce turbulence and increase flow. Removing it significantly alters flow characteristics.

CLOSE LIMIT MAIN JETS				
Basic Part Number 122-				
352	432	512	592	672
362	442	522	602	682
372	452	532	612	692
382	462	542	622	702
392	472	552	632	712
402	482	562	642	722
412	492	572	652	732
422	502	582	662	742

Close-limit jets are pretested to ensure that there is virtually no flow variation between jets of the same number. These jets were originally used only in laboratory testing but now are readily available. Flow variation among standard jets is small, so use of close-limit jets usually is restricted to special applications where mixture distribution, fuel economy, or emissions are being fine-tuned. Note that the first two digits of the close-limit jet number refer to the same jet size as the corresponding standard jet number.

HOLLEY STANDARD MAIN JETS			
Jet No.	Drill Size	Jet No.	Drill Size
40	.040	71	.076
41	.041	72	.079
42	.042	73	.079
43	.043	74	.081
44	.044	75	.082
45	.045	76	.084
47	.047	77	.086
48	.048	78	.089
49	.048	79	.091
50	.049	80	.093
51	.050	81	.093
52	.052	82	.093
53	.052	83	.094
54	.053	84	.099
55	.054	85	.100
56	.055	86	.101
57	.056	87	.103
58	.057	88	.104
59	.058	89	.104
60	.060	90	.104
61	.060	91	.105
62	.061	92	.105
63	.062	93	.105
64	.064	94	.108
65	.065	95	.118
66	.066	96	.118
67	.068	97	.125
68	.069	98	.125
69	.070	99	.125
70	.073	100	.128

Holley main metering jets are available in a wide range of size. Note that the "jet number" does not correspond to the orifice (drill) size. Also, in some cases the drill size of consecutive jets may be the same, but the jet with the larger number will always flow more fuel (because of other design changes).

Close-limit main jets are also available. They were originally developed for precise "emissions-type" metering, but are frequently used in race carburetors to achieve a higher degree of fuel metering precision. The flow variation of close limit jets has been reduced by approximately 60%, as compared to the 122- standard jet. Close-limit jets may be identified by a three digit number stamped on the side. The third digit may be an inverted "1," "2," or "3" and pertains to the position of the jet within the flow range. A "1" places the jet on the lean side; a "2" indicates middle of the range; a "3" places the jet on the rich side. The close-limit jets were originally available, by number, from 352 to 742. This corresponds to standard jets within the range of 35 to 74. It wasn't until 1980 that "1" and "3" suffix jets were released for general sale, and it may be necessary to purchase a jet kit to obtain close-limit jets of a desired size and limit range.

Super Tuning and Modifying Holley Carburetors

HOLLEY SECONDARY METERING PLATES

Main Hole	Idle Hole	Order By Part No.	Part Stamped
.052	.026	134-7	7
.052	.029	134-34	34
.055	.026	134-3	3
.059	.026	134-4	4
.059	.029	134-32	32
.059	.035	134-40	40
.063	.026	134-5	5
.064	.028	134-18	15
.064	.029	134-30	30
.064	.031	134-13	13
.064	.043	134-33	33
.067	.026	134-8	8
.067	.028	134-23	23
.067	.029	134-16	16
.067	.031	134-9	9
.067	.035	134-36	36
.070	.026	134-6	6
.070	.028	134-19	19
.070	.031	134-20	20
.070	.033	134-41	41
.071	.029	134-35	35
.073	.029	134-39	39
.073	.031	134-37	37
.073	.040	134-17	17
.076	.026	134-10	10
.076	.028	134-22	22
.076	.029	134-43	43
.076	.031	134-12	12
.076	.035	134-53	3
.076	.040	134-28	28
.078	.029	134-38	38
.078	.040	134-52	52
.079	.031	134-11	11
.079	.035	134-24	24
.081	.029	134-44	44
.081	.033	134-49	49
.081	.040	134-21	21
.081	.052	134-31	31
.081	.053	134-29	29
.082	.031	134-46	46
.086	.043	134-25	25
.089	.031	134-47	47
.089	.037	134-55	5
.089	.040	134-27	27
.089	.043	134-26	26
.093	.040	134-54	4
.094	.070	134-15	15
.096	.031	134-50	50
.096	.040	134-45	45
.098	.070	134-14	14
.113	.026	134-42	42

Holley offers a number of secondary metering plates for Model 4160 and 4175 carburetors. When switching plates, care must be taken to match idle feed diameter (idle hole) as well as main metering restriction size to particular requirements. An ID number is stamped on the face of the plate.

On Holley Model 4160 carbs, the metering of secondary fuel is controlled by a metering plate, not a metering block with replaceable jets. The metering plate also is found inside the easily removed float bowl and is held to the main body by six clutch-head screws. Metering orifices are drilled in the bottom of the plate (arrows), and the entire plate must be replaced to vary fuel metering response.

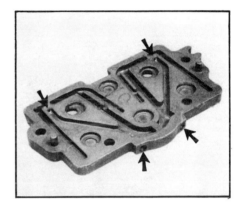

A backside view of a metering plate shows the fuel tracks and the four fuel metering restrictions. The lower arrows point to the main metering restrictions, the upper arrows point to the idle feed restrictions.

In the days of yore, before the oil companies discovered the utility of "shortages" as price-increasing devices, it was common practice to jet a carburetor slightly on the rich side. Now that fuel economy and low exhaust emissions are important considerations, the jetting pendulum has swung to the lean side. Generally, emissions-design carburetors can be installed in "out-of-the-box" condition; they are jetted relatively lean. High-performance carburetors, especially double-pumpers, are typically jetted relatively rich; for street use, jet size can usually be reduced a few numbers. The ideal jetting is as lean as possible without creating surge or misfire. Trial-and-error testing is the best method of determining the optimum jet size for a particular engine/drivetrain combination.

When attempting to rejet the secondary side of a 4160 or 4175 carburetor, a slightly different approach is needed. As a means of reducing manufacturing costs, Holley fits these model carbs with a metering plate rather than a metering block. In order to rejet, it is necessary to either change the plate or drill the main metering restrictions. Irrespective of the method chosen, working with the 4160/4175 configuration is cumbersome. A plate is roughly four times the cost of a pair of jets, and whether it is changed or drilled, six additional screws must be removed and replaced each time a change is made.

If jet changing is an ongoing occurrence, a metering block with removable jets should be installed. Holley conversion kit no. 34-6 may be used on model 0-1850 and other carburetors with side-hung float bowls; part number 34-13 is applicable to model 0-3310 and other carburetors with center-hung float bowls.

If secondary metering is simply too rich or too lean, a one-time plate change will be considerably cheaper than a conversion kit. Inasmuch as both main and idle restrictions are also located in the plate, it will be necessary to check the part number listing to insure that the idle-feed diameter is not inadvertently changed. As an example, blocks numbered 134-6, -19, -20 and -41 all contain a .070-inch main restriction but the idle-feed restriction is .026-inch, .028-inch, .031-inch and .033-inch, respectively. If a carburetor was originally fitted with a -23 plate and it was necessary to slightly enrich the main circuit only, a -19 plate should be installed because it contains the same diameter (.028-inch) idle-feed hole as the original part and a .003-inch larger main restriction. Drilling the restrictions in the plate can create the same problems as when drilling jets oversize. Before altering the diameter of the holes, the entry chamfer should be examined and the plate rechamfered, if necessary, after drilling. A listing of secondary metering plates and restriction diameters may be found on this page.

Viewing the channel side of a secondary plate can shed an abundance of light on the secret pathways of various fuel circuits. The routing is virtually identical to that of a metering block but is easier to understand since

Super Tuning and Modifying Holley Carburetors

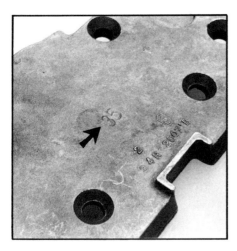

All Holley metering plates are stamped with an identifying number. As seen in the accompanying chart, this number 35 plate has a 0.071-inch main metering restriction and a 0.029-inch idle feed hole. Plates are available with several combinations of main and idle feed restrictions.

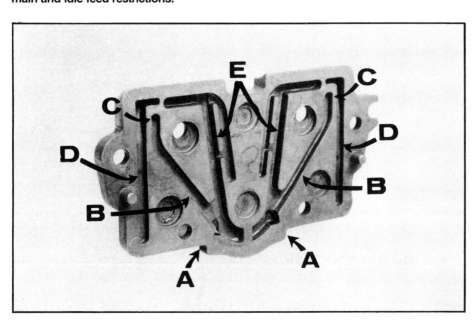

Fuel enters the metering plate through the main restriction (A). At idle, fuel moves up passages (B), through the idle feed holes (C), and down the down-well passages (D) to the idle transfer slot supply holes in the main body. When the secondary main circuit is activated, the fuel moves up passages (E). It is emulsified by bleed air supplied from the main body, and then it enters the connecting passages in the main body that lead directly to the secondary discharge nozzles.

The Holley power valve is an extremely important component in the overall fuel delivery picture. It is activated when manifold vacuum drops to a predetermined level as indicated by each valve's vacuum rating of the valve. A properly selected power valve will be closed at normal cruise to gain maximum economy and will open during low-vacuum/high-load conditions to provide the extra fuel needed for smooth power and acceleration.

all passages are readily visible. Beginning with the main metering restrictions (the equivalent of jets) where fuel enters at the bottom of the plate, flow proceeds either to the idle or main passages, depending on operating mode. At high engine speeds, fuel travels up the main well, where it is mixed with bleed air and into the main body passageway leading to the main discharge nozzle. Another passage branches off from the main well and leads to the idle feed restriction. When the secondary throttle plates are closed, fuel takes the outer branch, flows up the idle well and through the restriction. At this point it is emulsified and the mixture flows down the passageway that connects to the transfer slot supply hole in the main body. Since Model 4160/4175 carburetors, like the great majority of 4150s and 4165s, have no secondary idle-mixture adjustment, there is no separate idle circuit. All idle fuel is supplied either through the transfer-slot or the constant-feed hole situated directly below the slot (and supplied by the same circuit).

AIR BLEED MODIFICATIONS

As mentioned elsewhere, Holley engineers are largely opposed to air bleed modifications, except in rare competition-only applications where they recommend altering only the idle bleeds. Their position on the subject is understandable — altering bleed size is one of the easiest ways to ruin a perfectly good carburetor. Since drilling a hole larger is infinitely easier than making it smaller, the typical approach taken by most experimenters is to twist a wire drill through the bleed, opening it up a few thousandths. At this point a number of possibilities exist, most of which will not be music to the ears of the carburetor owner. With any amount of good fortune, the slight change in bleed diameter will cause nothing more than the need to rejet. But, it is entirely possible that an off-idle stumble will be present because the mixture is now leaner, and activation of the main system has been delayed. Richer jetting may minimize or eliminate the stumble. Then again, it may not. Surprising as it may seem, many racers, after discovering their bleed modifications have turned their carburetors into closet clutter, claim the carburetor is defective and attempt to have it replaced under warranty.

The real question is, what's to be gained by altering air bleed size? If the booster venturis have been changed, the resultant increase or decrease in signal strength may necessitate a bleed alteration as a means of reestablishing the original system timing. Such alterations require the use of a carburetor flow box, or a good deal of trial-and-error testing to verify that the fuel delivery curve has been properly tailored. In general, it's best to avoid making air bleed modifications unless experience is sufficient to assure some amount of success.

An exception to the caveat against air bleed modifications is the Model 4500 Dominator HP series. These carburetors are equipped with removable air bleeds which can be changed as easily as a main jet. This arrangement is conducive to experimentation and an "oops" can be easily rectified by simply reinstalling the original air bleed. Some Model 4150 carburetors sold by modification specialists are also equipped with removable air bleeds and these too are suitable for experimentation with air bleeds of various diameters.

POWER ENRICHMENT CIRCUIT

It would be difficult to estimate the number of racers and performance enthusiasts who set about changing power valves without the faintest notion as to the manifold vacuum conditions of the engine. The reason for such ignorance can usually be traced to the lack of a simple vacuum gauge. Such a device is the only external means of determining the internal condition of an engine. It is truly amazing that there never seems to be enough money to purchase a $15 gauge, even for a $30,000 race car.

Having climbed down off my soap box, I can now tell you that once you have a vacuum gauge and some insight into an engine's load profile, power valve changes will be much more useful. As mentioned in the previous chapter, Holley power valves are rated in terms of opening point, not flow capacity. All power valves of a given primary part number have the same fuel flow potential. A 125-XX valve will flow sufficient fuel to accommodate power valve channel restrictions only up to .090-inch. A 125-XXX high flow valve will handle PVCR's of .120-inch or smaller. Two-stage

Here the valve can be seen at the upper left, extending through the metering block above the main metering jets. Once the metering block is removed, shown on the right, the power valve can be removed by turning it counter-clockwise with a 1-inch wrench.

Holley's two-stage power valve was developed to increase part-throttle economy of heavily loaded recreational vehicles. By adding the first stage of enrichment at a relatively high manifold vacuum, it is generally possible to reduce the main jet size somewhat without encountering lean light-load stumbling. However, these valves should never be used in performance applications because they restrict overall fuel flow.

valves listed under part numbers 125-206 through 125-218 are intended for PVCR's of .060-inch or smaller.

Selection of a power valve opening point should be based on the amount of manifold vacuum present during "normal" operation. Obviously, "normal" will cover quite a range, depending upon the application in which an engine is used. A high-performance street vehicle may cruise at 18 inches and rarely see a reading of less than 6 inches (except during full throttle acceleration); an RV or motorhome will be lucky to cruise at 9 inches and will pull 18 inches only when rolling down a 10% grade loaded with

HOLLEY POWER VALVES

SINGLE STAGE POWER VALVES (includes gasket)

Standard Flow

Part Number	Opening Vacuum (in Inches of Hg.)
125-25	2.5
125-35	3.5
125-45	4.5
125-50	5.0
125-65	6.5
125-75	7.5
125-85	8.5
125-95	9.5
125-105	10.5

High Flow

Part Number	Opening Vacuum (in Inches of Hg.)
125-125	2.5
125-135	3.5
125-145	4.5
125-155	5.5
125-165	6.5
125-185	8.5
125-1005	10.5

TWO-STAGE POWER VALVES (includes gasket)

Part Number	1st Stage Opening (in Inches of Hg.)	2nd Stage Opening (in Inches of Hg.)
MODEL 4160		
125-206	12.5	5.5
125-207	10.5	5.0
125-208	10.5	5.5
125-213	11.5	5.0
MODEL 4175		
125-209	11	6.0
125-210	9	2.5
125-211	10.5	5.5
125-212	12	6.5
125-215	10	6.0
125-216	8	1.5
125-217	10	4.0
125-218	11	5.5
MODEL 4360		
125-200	9	5.0
125-201	8	5.0
125-202	8	4.0
125-203	8.5	5.5
125-204	8	3.0
125-205	9	3.0

kids, in-laws, dogs, a barbeque and two Mopeds.

The opening point of a power valve must, therefore, be correlated to vehicle use. Carburetors used on mild high-performance engines can usually be fitted with a 10.5-inch (as in inches

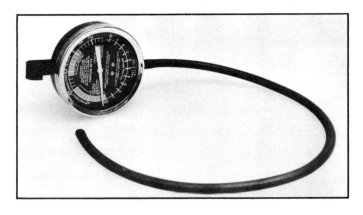

If you're going to custom-tailor the power valve opening point to suit a specific engine, a vacuum gauge is the best way to check the engine requirements. Engine vacuum should be checked at idle, normal cruise, and moderate-to-heavy acceleration. The opening point should be below the lowest reading during "normal" cruise conditions.

As a rule of thumb, or a rule of power valves in this case, jet size must be increased six to eight numbers to provide additional fuel if the power valve has been removed. However, the size of the PVCR varies considerably among different list numbers and eight jet sizes may not be sufficient to prevent a lean condition. The PVCR diameter should be checked and its area used as a gauge for compensatory jetting when the power valve is made inoperative.

of mercury) valve. The advantage of a 10.5-inch opening point is that fuel will start flowing through the power enrichment circuit more quickly when the accelerator pedal is pushed briskly toward the floor (because the valve isn't waiting until manifold vacuum drops to 8.5 in/Hg. before it opens). However, use of a valve with such a relatively high opening point is predicated upon the fact that manifold vacuum never drops below the valve's opening point during normal acceleration. If it does, the valve will needlessly pop open and flush a little more gas mileage down the drain. For precisely this reason, vehicles with low power-to-weight ratios typically require power valve opening to be delayed until vacuum drops below 6.5 inches.

In theory, the opening point should be as high as possible, but below the lowest point within the "normal cruise" range. Viewing the power valve as an "extra gas switch," it's obvious that the switch should only be thrown when an engine is placed in a low-vacuum situation for more than a second or two. Most production Holley four-barrel carburetors are fitted with a 6.5-inch or 8.5-inch valve (or a two-stage valve). Such ratings are satisfactory for most engines — the valves open soon enough to prevent a lean stumble when transitioning from

One of the keys to outstanding reliability and performance of a Holley carb is the simplicity and flexibility of the accelerator design. As the throttle is opened, the pump cam located on the backside of the throttle lever (A) pushes upward on the pump operating lever. This lever rotates around the fulcrum at (B) and pushes downward on the pump housing lever (C). As the pump housing lever pushes against the diaphragm inside the pump housing, fuel is forced through the delivery circuit to the accelerator pump discharge nozzle, where it is discharged into the incoming airstream.

cruise to full throttle or high-load conditions, yet late enough to prevent opening during light acceleration.

To get a better handle on power valve opening points, consider the requirements of a performance-oriented engine in a fairly typical street machine or street rod. At idle (no load), the manifold vacuum reading is about 15 inches. During normal road conditions (under load), the vacuum ranges from 10 inches (light acceleration) to 17 inches (highway cruise). Only during impromptu "speed contests" or when climbing steep hills will the vacuum drop below 10 inches. During normal acceleration away from a traffic light, manifold vacuum swings from 15 inches at idle

When the accelerator pump housing is removed, the pump diaphragm is exposed. This diaphragm must not be cracked or torn if the pump is to operate correctly.

down to 10 inches during initial acceleration and up to 19 inches as RPM increased. When the transmission is shifted into high gear the reduction in torque multiplication would increase engine load and decrease vacuum to 16 inches.

Consider these parameters:

1) A 10.5-inch valve would open unnecessarily during normal acceleration, causing a deterioration of gas mileage.

2) A 6.5-inch valve would leave a fuel delivery "hole" with no mixture enrichment from 6.5 inches to 10 inches. In the example above, any time vacuum drops below 10 inches, the engine is in a high-load condition. Therefore, delaying actuation of the power circuit until vacuum drops to 6.5 inches creates a transient condition that is conducive to stumble, lean misfire, and ping.

3) An 8.5-inch valve would reduce the size of the "hole," eliminating most, and probably all, of the problems associated with leanness.

4) A 9.5-inch valve would be ideal in theory, opening just 1/2-inch below the point at which enrichment was deemed necessary. In practice, it might prove ideal, or it could hurt fuel economy. Under "real life" conditions, a 9.5-inch valve might open at an indicated 10 inches on the gauge or vice versa (due to tolerances).

As in most things automotive, trial-and-error experimentation is the acid test. Unfortunately, with the continuing variations in gasoline quality, experimentation has become almost essential to the maintenance of performance and economy. In many instances it is necessary to initiate power enrichment before the optimum time simply as a means of preventing detonation during throttle transitions. While the engine might not be laboring noticeably, the load associated with an 8- to 10-inch vacuum reading is frequently enough to make a respectable gasoline engine clatter like a diesel.

It is under these conditions that a two-stage power valve can be of use. By bringing in a first stage of enrichment at 9-12.5 inches of vacuum, a lean cruise mixture can be maintained while surging, stumbling, and misfire under light load can be eliminated.

Depending upon the vehicle and the manner in which it is driven, it is often possible to reduce the main jet size somewhat when a two-stage power valve is installed. It should be noted, as previously mentioned, that use of the two-stage valve should be limited to metering blocks with PVCR's measuring .060-inch or less. For vehicles used in competition, installation of these valves is not recommended. Since many high-performance (and all race) engines are fitted with long-duration camshafts, vacuum at idle may not be sufficient to keep the first stage closed. Also, as vacuum climbs with increasing RPM, it may reach the point where it causes the second stage to close, causing a high-speed leanout. This is the predicament that race engine tuners face when attempting to choose a single-stage valve.

With an engine that idles at very low vacuum (2.5-4.5 in/Hg), a power valve that remains closed at idle may return to the closed position when vacuum increases at high engine speeds. Some racers have chased a variety of phantom ignition, fuel pump, or camshaft problems because a car "laid down at the top of each gear" or toward the end of a long straight. Actually, what they were experiencing was power valve closure. The cure is a valve with a higher opening point.

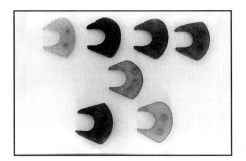

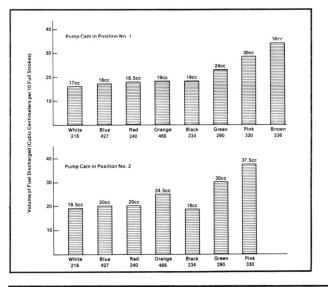

Considerable tuning of the accelerator pump circuit is possible because of its cam actuation. Several interchangeable pump cams are offered, each with a different fuel delivery curve.

An important accelerator pump characteristics is the total volume of fuel delivered during pump operation. This chart compares total volume supplied (for every 10 strokes of the pump) with each different cam. Note that volume delivered varies with mounting position.

Degrees of Throttle Opening	290		240		427		466		234		330*		
	1	2	1	2	1	2	1	2	1	2	1	2	3
-0-	-0-	-0-	-0-	-0-	-0-	-0-	-0-	-0-	.0175	.0200	.0275	-0-	-0-
5	.045	.050	.025	.025	.0325	.045	.0225	.0275	.030	.032	.045		.020
10	.0775	.085	.045	.045	.045	.075	.045	.050	.045	.045	.0625		.035
15	.100	.110	.060	.0625	.075	.100	.0625	.0675	.055	.057	.075		.045
20	.120	.130	.0775	.0775	.080	.1225	.080	.085	.065	.070	.085		.055
25	.1325	.150	.090			.135	.095	.100	.075	.080	.095		.0625
30	.1425	.1625	.1075			.1425	.1075	.115	.085	.090	.1025		.0725
35	.150	.170	.1075				.1175	.1225	.090	.097	.1125		.0825
40	.155	.175	.1125				.1225	.130	.097	.105	.1250		.095
45			.115				.1250	.135	.100	.110	.135		
50							.1275	.140	.105	.120	.145		
55											.155		.120
60											.160		
65											.1675		
70													
75													
80	.155	.175	.115	.0775	.080	.1425	.1275	.140	.105	.120	.1725		.140

* Note that cam 330 has three mounting holes.

This accelerator pump cam chart shows a comparison of the relative lift of each Holley pump cam. Each cam is identified by a different color and a three-digit code stamped on the side. In most cases, careful experimentation will be required to select the optimum cam profile for a specific racing application.

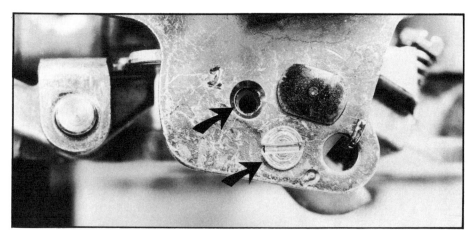

The mounting position of the cam affects pump-shot duration and total discharge volume. Position 2 generally provides increased lift and duration (depending on cam profile).

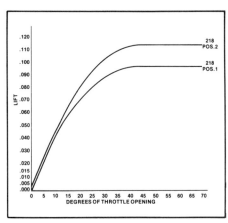

This graph shows the effect of the mounting position on total lift. An example is the white "218" cam used on most stock Holley models.

In some instances it may be impossible to find a power valve that will remain closed at idle, yet remain open during high-speed operation. Occasionally the extra richness caused by the valve being open at idle may not present problems with fouled spark plugs or exceptionally erratic engine operation. Given enough camshaft duration, an extra shake or shudder will never be noticed. Alternately, increasing idle speed a few hundred RPM may bring vacuum up enough to alleviate an otherwise problematical situation. If all else fails the power valve can be removed and replaced with a plug. When this is done, jet size will have to be increased accordingly; rejetting should be based on the power valve channel restriction areas that are no longer being used to deliver fuel.

It would appear that the change in air/fuel ratio caused by an open power valve, or increased jet size, should have virtually no effect on the idle circuit, which receives fuel through an orifice that is less than half the diameter (and less than one-quarter of the area) of the main jet. However, the extra fuel in the main well does tend to increase richness of the idle mixture. It can also cause a very rich off-idle condition. Typically, altering jet size by a few numbers will have little effect, but increasing jet size by six or more numbers, (as is required when removing the power valve), or having the power valve open at an inopportune time, can create a noticeable difference. In certain instances, after the main metering/power enrichment relationship is firmly established, it may become necessary to reduce the diameter of the idle feed restrictions in order to gain a clean idle quality.

While extensive power valve experimentation is usually necessary only with fully modified race engines, the need for alteration of PVCR diameter can occur over a much wider range of applications. Generally, recalibration of the power enrichment circuit is required when a carburetor is used in

The easiest method of adjusting the pump override screw is to compress the spring until free play between the screw and lever is evident. Then back the screw off until all play is removed.

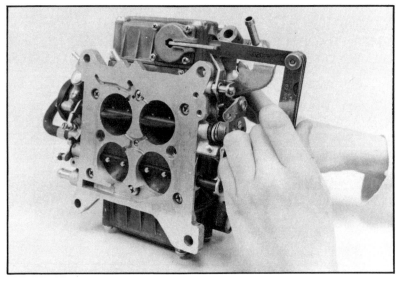

The traditional "engineer's method" of accelerator pump override screw adjustment is to hold the throttle wide open and check for 0.015- to 0.20-inch clearance (to ensure that the pump isn't binding). However, many people do not read the Holley installation instructions carefully and adjust this clearance with the throttle in the idle position. This results in poor off-idle performance.

an environment for which it was not originally intended. Many race-type carburetors have relatively small PVCR's because it is assumed that jetting will favor the rich side and that power enrichment need not be excessive. If these carbs are jetted considerably leaner for street use it is possible (and recommended) to drill the PVCR's to a larger diameter in order to obtain a proper full-power mixture. In the event that a PVCR must be reduced in size, it may be drilled oversize, plugged with a small insert and redrilled to the desired size.

ACCELERATOR PUMP CIRCUIT

The unique method of actuating the Holley accelerator pump with a cam, rather than a lever-and-rod arrangement, provides a number of tuning opportunities not available with other carburetors. By varying the shape of the cam and adjusting cam position with respect to the throttle lever, both pump timing and discharge volume can be altered. Each cam is color-coded and has a number stamped in its side. A complete assortment is available (part no. 20-12).

The "white" cam, number "218," is probably the most commonly used piece. It is found on performance as well as emissions/design models. In general, it will be difficult to bring about dramatic improvements in accelerator pump operation simply by changing the cam. As is the case with the eccentrically-shaped device that operates the engine intake and exhaust valves, the accelerator pump cam is a timing device. Altering the profile of the pump cam changes accelerator-pump fuel-delivery timing with respect to the throttle-opening cycle, causing a minor change in the volume of fuel discharged at the pump nozzle.

With a street-driven vehicle, cam lift should increase at a slow rate relative to throttle opening. Just enough fuel must be discharged early in the throttle opening cycle to eliminate an off-idle stumble, yet there must be sufficient lift remaining for part- to full-throttle transitions. Conversely, in a drag racing application the greatest volume of fuel is normally required at one specific point (usually later in the

The primary accelerator pump discharge nozzle is located inside the choke shroud (in this case the choke plate has been removed to show the location of the nozzle). On vacuum-actuated models there is only one nozzle, and it is located as shown in (A). The model shown here is a double-pumper. The secondary accelerator pump discharge nozzle is located as shown (B).

throttle cycle). A drag race car needs help from the accelerator pump only when leaving the starting line. With an automatic transmission and torque converter the starting line RPM may be low enough that a small engine with a large carburetor will stumble when the carb is sharply brought to wide-open throttle. This was a problem in the days of 3000 RPM stall speeds, but in recent years torque converter technology has raised stall speeds to the point where the main system is feeding fuel while the car is staged. Once a significant amount of fuel has begun to flow through the main metering circuit, part- to full-throttle transition requires little supplemental fuel.

With a standard-shift drag car that leaves the starting line at 8000 + RPM, the accelerator pump circuit should be tailored to ease driving through the pits and up to the starting line. The same can be said of a high-speed oval-track car where changes in throttle setting (going into and coming out of the corners) are usually minimal. Short track (1/4-1/2 mile), figure-8 and road-racing competition require rather dramatic movements of the accelerator pedal and the equipment to which it attaches. In this sense, driving on the race course becomes analogous to motoring

Accelerator pump discharge nozzles, also called "shooters," are available with different size discharge orifices. The nozzles are marked, as shown, to indicate the orifice size (see chart in Appendix). The size of the nozzle and the pump cam shape work together to coordinate pump "shot" delivery volume during various portions of throttle lever movement. The shot can be adjusted to be long or short, early or late, in the throttle cycle.

down the highway, so many of the same considerations apply.

Marine engines typically operate under heavier loads than their land-based counterparts and consequently may require a more aggressive accelerator pump discharge rate. Another factor affecting accelerator pump activation in marine engines is cam duration. Many of the engines found in pleasure and ski boats are blessed with raspy camshafts that generate relatively low amounts of manifold vacuum. Under these conditions, when the throttle is rapidly opened, it opens an unusually large "hole" in the fuel delivery curve. Consequently, more aggressive accelerator pump actuation is required.

Depending on the pump cam, the cam mounting position (there are two) and the volume inside the pump housing, pump discharge volume can vary from approximately 17 to 37

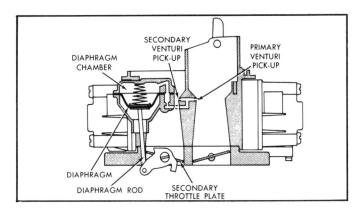

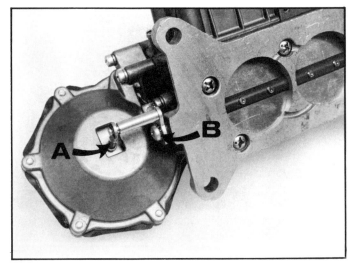

Although they're called "vacuum-actuated" secondaries, normal manifold vacuum does not open the secondary throttles. During high-speed/heavy-load conditions, manifold vacuum is low but airflow velocity through the primary venturis is high. This airflow is used to create a vacuum signal at the primary venturi pickup. The secondary venturi pickup is much smaller and acts to bleed off some of the vacuum signal until the primary airflow is sufficient to require secondary activation. As airflow begins in the secondary venturi, the secondary pickup also begins to create vacuum, increasing the rate of secondary activation.

As the diaphragm is pulled upward by venturi vacuum, the diaphragm operating rod (A) rotates the throttle shaft (B) toward the open position. The example shown is a vacuum-actuated Model 2300 and is also typical of vacuum-secondary 4150/4160-series Holleys.

The secondary diaphragm spring is located above the vacuum diaphragm. It holds the secondary throttles closed at idle and returns the throttles to the closed position when air velocity and venturi vacuum decline.

cubic centimeters per 10 full pump strokes. This means that by simply changing the pump cam and/or housing may alter the volume of fuel discharged by as much as two cc's. Being able to more than double fuel delivery by simply swapping pump cams provides for a great deal of tuning capability. It is also very easy to supply too much fuel through the accelerator pump circuit. Always remember that, regardless of the application, the pump circuit should provide just enough fuel to insure against a stumble. Once this point is reached, additional fuel will degrade performance as the engine will have to recover (sometimes taking up to a second or two) from an over-rich condition.

Once the optimum cam profile is selected, through the trusty trial-and-error (or is it trial by ordeal) process, the spring-loaded pump lever operating screw must be properly adjusted. Existing literature, which references a .015-inch clearance, has led to fairly widespread misunderstanding concerning the correct adjustment procedure. There should be no clearance between the operating screw and pump lever when the throttle shaft is in the idle position. The referenced .015" clearance — actually a minimum; maximum recommended clearance is .062" — pertains to measurements taken when the throttle shaft is moved to the wide-open throttle position and the pump lever is manually pushed as far down as it will go.

In actual practice, this method of adjustment is rather cumbersome. The intent of adjusting the lever screw is simply to eliminate all clearance at idle, while avoiding compression of the pump diaphragm. This may be easily accomplished by tightening the adjusting screw-and-nut against the override spring until a clearance at idle is present, then loosening the nut gradually until all clearance is removed and the slightest of preloads exists. After this, there will almost always be sufficient clearance at wide-open throttle to prevent binding.

This adjustment affects both timing and volume of fuel discharged by the pump circuit. If clearance is present when the throttle lever is at idle, pump action will be delayed until that clearance is taken up (by opening the throttle). At the other end of the adjustment spectrum, preloading the operating lever (by loosening the screw-and-nut assembly too far) will provide an immediate fuel discharge at the pump nozzle, but total volume will be reduced. If it is impossible to obtain proper adjustment with the screw, an incorrect cam or lever assembly is probably at fault.

One of the problems that frequently interferes with selection of the appropriate cam profile centers around discharge-nozzle diameter. By varying the size of this nozzle, it is possible to influence the rate at which fuel is discharged. Decreasing nozzle size restricts fuel flow; accordingly, with the same cam, discharge duration is lengthened. In some cases, the nozzle can be so restricting that the pump override spring will be compressed because cam lift is increasing at a faster rate than fuel can be discharged. Since gasoline is not compressible, something has to give. The override spring compensates for this condition and is, in fact, included in the pump assembly to prevent linkage damage in such cases.

Irrespective of nozzle diameter, an identical amount of fuel will be discharged for a given amount of cam lift, but with a restrictive nozzle, fuel flow will continue after throttle shaft movement stops. The residual discharge is the volume of fuel that caused compression of the override spring. However, if fuel is still flowing from the pump nozzle after throttle shaft movement has ceased, it is of no real use to the engine. Lengthening discharge duration is typically useful only with extremely heavy vehicles with high rear axle ratios. In such cases, engine RPM under load increases very slowly, so it takes longer to activate the main system. A longer pump shot is needed to prevent the engine from encountering a lean condition prior to main system startup. Race cars present an entirely different situation. RPM rises quickly and maximum fuel volume is usually needed as soon as the throttle is moved toward wide open.

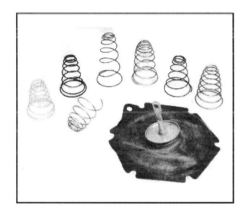

The point at which the secondary throttle plates are activated can be adjusted by removing and replacing the diaphragm counterbalance spring. Several springs are available in Holley kit 20-13.

To ease the chore of changing secondary diaphragm springs, Holley developed a quick-change diaphragm housing cover (part number 20-59). It eliminates the need to remove the whole diaphragm housing when changing springs.

In addition to nozzles of different sizes, Holley also offers three different nozzle styles. The standard nozzle is a simple casting with two drilled orifices. A second type consists of the same casting but includes a tube pressed into each discharge hole. These tubes direct the "pump shot" closer to the center of the booster venturi. The third version is an "anti-pullover" design necessitated by the location of the check valve in 4165/4175-type metering blocks. Rather than a needle valve positioned just below the nozzle screw, the 4165/4175 Spread-Bore carburetors use metering blocks with a ball-type check valve at the bottom of the diagonal passageway that delivers fuel to the main body. Since a significant amount of fuel is located between the check ball and the nozzle itself, an anti-pullover squirter must be used to prevent fuel siphoning at high air flow. An anti-pull-over squirter can be installed in a standard 4150/4160, but offers no advantage on standard applications.

In some race applications it is possible to use the siphoning tendency to an advantage. By removing very small amounts of material from the pump-discharge needle valve (4150/4160 only) weight can be reduced sufficiently so that very high airflow will lift the needle, allowing fuel to be drawn out of the nozzle. This fuel then serves to provide auxiliary high-speed enrichment. However, weight removal should be done very judiciously or the accelerator pump circuit will begin siphoning even during low airflow. A standard check needle weighs only about 1-1/2 grams, so there isn't a lot of margin for error when removing weight. Such highly specialized modifications are best left until all other problems have been sorted out and the car is running consistently. A detailed description of accelerator pump modifications for race use is contained in chapter 8.

VACUUM-ACTIVATED SECONDARIES

Carburetor manufacturers have spent thousands of dollars designing systems that will open the secondary throttles smoothly and positively. But it only takes a racer about 30 seconds to decide that such systems are junk

Use of dual four-barrels necessitates that linkage be properly adjusted so that both carburetors reach wide-open simultaneously. When vacuum secondaries are used, diaphragms must be balanced to avoid fuel-air distribution irregularities. The vacuum diaphragm covers included in Holley part number 20-73 contain balance tubes that can be connected to each other.

Many double-pumpers use a small link to actuate the secondary throttle plates. The link can be bent slightly to adjust the overall length, ensuring that primary and secondary throttles reach full open position.

Super Tuning and Modifying Holley Carburetors **53**

HOLLEY NEEDLES & SEATS		
VITON INLET		
Seat Size	Order By Part No.	Type
.097"	6-506	Adjustable
.097"	6-507	Adjustable
.097"	6-508	Adjustable
.097"	6-517	Adjustable
.110"	6-504	Adjustable
.120"	6-518	Adjustable
.101"	6-520	Adjustable
2mm	6-512	Model 5200
.0785"	6-511	Non-Adjustable
.100"	6-516	Non-Adjustable
.110"	6-510	Non-Adjustable
.110"	6-514	Model 4360
.97"	6-513	Off-Road
STEEL INLET		
.97"	6-501	Adjustable
.110"	6-500	Adjustable
.120"	6-502	Adjustable
.130"	6-515	Adjustable
.150"	6-519	Adjustable

Holley needle-and-seat assemblies are easily replaced. However, it is seldom necessary to change to a larger size. Holley racing carbs already are fitted with high-flow assemblies, and installing a larger-than-stock assembly on a street engine is generally a waste of time. The Viton assemblies are preferred for all applications, except when the carb will be used with non-gasoline-based fuels.

— even for street applications. Not true. The only problem with vacuum secondaries is that they open too slowly for most performance enthusiasts' liking. But, by simply changing the spring within the diaphragm housing, the opening rate can be tuned to virtually any requirement. Assuming, of course, that the diaphragm has not been punctured. If it has, it must be replaced with a new one. There are five part numbers for secondary diaphragms, so it is important that new and old match. The difference between diaphragm assemblies is usually the length of the throttle lever connecting rod.

Holley kit number 20-13 contains a number of color-coded secondary diaphragm springs. By changing the spring, the opening rate can be optimized for any engine/chassis configuration. Heavy vehicles with small engines will require a relatively stiff spring in order to prevent the secondaries from opening too soon and causing a stumble. Lighter cars with larger engines will tolerate a much lighter spring load because RPM increases more rapidly.

The Holley inlet needle-and-seat valve is a rugged and reliable assembly. A Viton-tipped needle is used in most box-stock carbs to improve resistance against small impurities in the fuel that affect fuel inlet control. The needle-and-seat controls the all-important fuel level in the float bowls, and the assembly should be checked or replaced at normal rebuild intervals to ensure proper operation.

Conversion to mechanical actuation is a mistake since there will be no accelerator pump circuit supplying fuel to the secondary bores. As is the case with too weak a spring, mechanical actuation can result in a severe stumble and misfire.

One pitfall to beware of when experimenting with vacuum secondary spring is a "seat-of-the-pants" feel. If it isn't moderated by some "brain-in-the-head" thought, performance will not be consistent. The feel of the secondaries "kicking in" is greatest when there is a slight hesitation preceding the power surge that accompanies a "foot-to-the-floor" movement. In fact, what is occurring is that the secondary throttle plates are opening too quickly, causing a decrease in the rate of acceleration prior to the "surge." As opposed to a stumble, which is quite noticeable, a hesitation is barely perceptible. Frequently, the driver misinterprets his "seat-of-the-pants" feedback. The feel of acceleration may be greater when the hesitation is present, but the vehicle is slowing down for a split second before it speeds up.

When tuning the secondary spring, all testing should be done in high gear to maximize engine load. If there is no stumble in high, there won't be any in the lower gears. But, with additional torque multiplication, an opening rate that is satisfactory when the transmis-

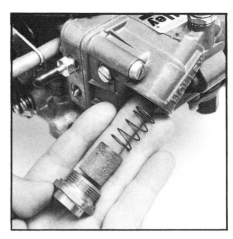

Holley carbs are fitted with a fuel filter inside the inlet fitting (the model shown here is a 4150 with center inlet bowl). It is often recommended that these filters be removed and discarded for performance use. Don't believe it. For all-out racing this may be a good idea, but only if a separate, high-volume filter is installed elsewhere in the fuel delivery system.

A reliable filter must be used in every fuel system. In addition to the stock Holley inlet filter, it is a good idea to use a separate replaceable-element, high-capacity fuel filter in the fuel line leading from the fuel pump to the carb inlet(s).

sion is in first or second gear may result in problems when accelerating rapidly in high gear.

MECHANICALLY ACTUATED SECONDARIES

Over the years, Holley has employed several mechanical secondary linkage arrangements. The latest version as used on many double-pumpers consists of a short link which interconnects the primary and secondary throttle levers. An oblong hole in the secondary lever allows the primary throttle to be opened a predetermined amount before secondary activation. This arrangement is very simple and virtually foolproof. The only adjustment, as such, is to check the action of the interconnecting link and bend it as necessary to insure that the secondary throttles reach full open when the primary throttles are wide open. This may be most easily accomplished by having someone sit in the vehicle and push the accelerator pedal to the floor while a second person monitors carburetor operation. (It is advisable to perform this check before the float bowls are filled with fuel, since this will eliminate fuel loss through the accelerator pump each time the throttle is moved.) When the final link adjustment is made and the throttle return spring is installed, the engine can be started to verify all other aspects of carburetor operation. After shutting the engine down, a double-check can be made to insure proper link adjustment.

Earlier versions of some double-pumpers (usually a plain four-digit part number or a "-1" revision) use a more intricate secondary linkage that pivots on a main body boss and is activated by a roller running inside a cam slot on the primary throttle lever. Adjustment of this "overhead" linkage may be done by either bending the secondary link rod or modifying the cam slot.

All Model 4500 carburetors are fitted with a roller-and-cam-slot secondary linkage. As opposed to the 4150 series carburetors, the linkage is internal, tucked between the primary and secondary throttle bores. Three cam-slot profiles are available for direct (1:1), staged, or "soft-staged" operation. These linkages are described in detail in Chapter 9.

FUEL INLET SYSTEM

Without an adequate fuel supply the best carburetor in the world cannot function. After all, of what use is a complement of fuel metering circuits where there is an insufficient amount of fuel to meter? In response to this question many enthusiastic carb tuners install the largest needle-and-seat assembly available, eliminate every filter and baffle in the system, and then wonder why a bad problem has become worse.

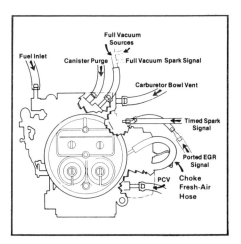

Vacuum leaks are the most common cause of carburetor problems. During any installation, all sources of potential leaks should be carefully checked. It is also important to understand the difference between "manifold vacuum" and "ported vacuum." On many late-model carb designs, ported-vacuum is used for emissions-related equipment and does not provide a vacuum signal at idle. It becomes operational only after air velocity is established in the primary venturis (off-idle conditions).

In general, very little in the way of modification need be done to any Holley fuel inlet system. At the very least, each stock bowl is fitted with a .097-inch needle-and-seat, and most carbs calibrated for high-performance are fitted with .110-inch inlet valves. With this type of inlet capacity, any problems with insufficient fuel flow are usually to be found elsewhere in the fuel delivery system. Only if fuel level is set too low, or if the inlet filters become clogged, or if a piece of dirt becomes lodged in the inlet seat, can a fuel supply inadequacy be caused by the inlet mechanism. More commonly, the impediment lies in the gas tank, fuel line, or fuel pump(s).

For typical high-performance use, installation of .110-inch needles-and-seats and setting fuel level so that it just dribbles out when the sight plug is removed (or setting internally adjusted floats to the recommended level) should provide trouble-free operation. On carburetors equipped with side inlet float bowls, it may be desirable to install center inlet bowls for improved fuel control during hard cornering. For competition use with some Model 4165 carburetors, conversion to dual feed bowls (kit no. 34-4 for GM applications; 34-5 for Chrysler applications) will make minor fuel level

During the summer months, fuel may percolate in the float bowls. An extra-thick manifold gasket may be installed to help combat this problem. It is available from Holley as part number 108-12 and is supplied with extra-long studs. If you use this gasket, do not tighten the carb hold-down fasteners too tightly because you may warp or break the throttle body.

adjustments easier. These bowls have externally adjustable floats.

Use of Nitrophyl floats is sometimes favored for special applications. These floats reputedly have greater sensitivity to fuel level changes, but Holley engineers question this so-called advantage. The question of float sensitivity is really more theoretical than practical. With the exception of alcohol applications, where all fuel delivery systems are stretched to their limits, additional float sensitivity will probably not solve many problems.

A number of years ago, Holley began using floats made of Duracon (a plastic material) rather than brass, in most production carburetors. As might be expected, Duracon floats are less expensive to manufacture and may offer better durability. However, Duracon floats are considerably lighter than their brass or nitrophyl counterparts and therefore ride higher on the fuel in the float bowl. When dry-setting Duracon floats, a slightly higher level is required to establish the same float bowl fuel volume. Dry setting of float levels should be used as a starting point; wet setting (with the float bowls full of fuel) is far more accurate because it reflects actual rather than theoretical fuel levels.

Super Tuning and Modifying Holley Carburetors **55**

It may seem strange, but for maximum efficiency, the air going into the carb should be cool. For maximum performance on racing and special-purpose street machines, a "cold-air" induction system can provide a noticeable increase in power.

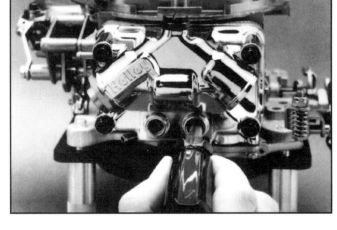

Jet changing is simplified considerably by the installation of Quick Change fuel bowls. Removal of two plugs in the front of the float bowl provides access to the main jets, eliminating the need to remove the float bowls to perform a change.

Besides float levels, needle-and-seat assembly selection must also be considered as a means of optimizing performance. Converting from Viton to steel needle-and-seat assemblies can solve problems when exotic fuels or gas additives are used. The chemicals in these compounds attack the Viton rubber tips used on standard Holley needles. Common sizes and styles of steel replacements are listed in the accompanying chart. If the float-adjusting screw-and-nut has become worn, a replacement set may be ordered under part number 34-7.

TUNING FOR HIGH ALTITUDE

It's no secret that air density decreases as altitude or temperature increases. However, what is not commonly known is that Holley carburetors are calibrated for conditions at sea level (everything near Detroit is at sea level) and 70°F (they must use heaters in the calibration lab).

As a rule of thumb, in order to maintain the same air/fuel calibration as was present at standard temperature and barometric pressure (70°F and sea level), jet size should be dropped one number (or .002-inch) for each 35° increase in temperature, or 2,000 foot increase in altitude. Since most of the inhabited portions of this country lie between sea level and 1500 feet, where 70°F temperatures are not uncommon, compensatory jetting for altitude or temperature extremes is generally not required.

For extended use at elevated altitude, fuel economy and performance can be optimized by rejetting. Whenever this is done the power valve(s) should also be replaced, as engine vacuum also decreases as you drive to new heights. Each time the altimeter is bumped by 3000 feet, power valve activation should be delayed by 1.5-2 in/Hg. (For example, a carb equipped with a number 85 power valve will typically deliver better fuel economy at 3000 feet if fitted with a 65 valve. Since altitude has reduced engine vacuum, this lower setting should not cause excessive leanness when transitioning to a heavy load condition.)

Rejetting for a quick trip to the mountains is ill-advised. If the high altitude jetting is not corrected when returning to the low lands, the excessively lean mixture could result in engine damage. On the other hand, anyone who lives in an extremely cold climate may want to consider richer jetting as a means of improving drive-ability during the long, hard winter.

GENERAL TUNING TIPS

FILTERS

Too frequently people who should know better lose sight of the fact that dirt and the accursed foreign matter so frequently written about, can wreak havoc with an otherwise perfectly adequate fuel system. Racers are especially vulnerable to dirt-induced problems. Often, gas tanks from wrecked cars are used in a racing chassis or a "custom" gas tank is made using improper materials or techniques. It seems strange that a drag race car that may be driven 200 miles a year has fuel system difficulties never encountered with a street-driven vehicle. To prevent unnecessary dirt-provoked fuel problems:

1) Always run a filter of some sort, even if it is just the stock brass element in the fuel bowl inlet. For high-flow applications use a Holley universal inline filter, or a Fram HPG-1 fuel filter. Both have low-restriction replacement filter elements.

2) Inspect the vehicle gas tank periodically to insure that internal coatings or galvanizing has not begun to chip.

3) When using electric fuel pumps, inspect periodically for accumulations of dirt and/or rust.

4) Flush all fuel lines before installation or after rework.

5) When installing a fuel-flow measuring device (as is usually included in an onboard gas mileage computer) insure that it is of sufficient flow capacity to accommodate full throttle fuel requirements.

6) When allowing a car to sit for extended time periods, flush tank and

lines before refilling carburetor float bowls. Drain gas tank and check for water in fuel.

7) Drain and remove fuel bowls and clean unwanted sediment.

8) Add a can of "dry gas" at least twice a year to eliminate water which may ultimately cause fuel system corrosion.

9) Inspect the air filter periodically and replace as necessary. A dirty filter not only reduces economy and performance but the proximity of dirt accumulations to the air horn can lead to a plugged air bleed.

10) Spray all air bleeds periodically with a carburetor cleaner, such as Gumout, to remove dirt and fuel "gum" accumulations.

CARB HEAT

An enigma of carburetor operation is that fuel should be cool when in a liquid state but it must be heated for satisfactory vaporization. To this end several techniques are employed to keep the carburetor and inlet fuel as cool as possible, while heat is applied either to the intake manifold, incoming air, or both.

For racing applications, it is common practice to block the heat riser passages in the intake manifold or use a manifold and/or cylinder heads with no plenum heat provision. The cooling effect of this practice can improve horsepower by a few percent and is useful in competition where every little bit helps. For street-driven applications, the blocked heat riser can be more harm than benefit.

A horsepower gain is still available, but at the cost of fuel economy, driveability, and engine life. Without exhaust heat at the manifold, vaporization is inadequate until the engine reaches operating temperature. In addition to operational problems (which are not really all that difficult to live with), excessive amounts of liquid fuel in the intake manifold can reach the cylinder walls and wash away the oil that keeps rings and pistons happy. This may not be a problem in a race engine where throttle movement is minimal when the engine is cold. But with the on-off-on cycles that are required when in traffic, or when pulling away from stoplights, the frequent shots of non-vaporized accelerator-pump fuel can lead to rapid ring wear.

In nearly all late-model vehicles, the incoming air is also preheated by the exhaust system. This is a further aid to vaporization and, for emissions purposes, enables an engine to run satisfactorily on a leaner air/fuel mixture. Ideally an engine should receive warm air when in cruise mode and cold air when under full power. A high-performance engine with a vacuum-controlled cold-air intake is an ideal arrangement for street use. To achieve optimum performance in such cases, it will be necessary to calibrate main metering for the warm underhood air and increase flow through the power enrichment circuit to supply the additional fuel necessary for operation with cold air at full power.

Along with whatever air supply system is used, it may be desirable to cool the carburetor to minimize fuel percolation caused by high engine-compartment temperatures. A thick insulating gasket between carb and manifold will help. If percolation persists, it may be necessary to install ducts or diffusers to deliver a flow of air past the carb exterior.

COLD

With all the plumbing that festoons emissions-era engine compartments, problems caused by extreme cold are rare. However, when temperatures are between 30°-50°F and humidity is high, carburetor icing can occur during engine warmup. Icing of an automotive carburetor is almost always associated with low engine speeds (idle) because the gap between the throttle plate and bore is minimal. Water vapor in the air condenses, freezes, and blocks the gap, causing the engine to stall. Once normal operating temperature is attained, there is sufficient heat at the plate/bore junction to prevent icing. With a properly functioning choke mechanism (with fast-idle provision) and the availability of manifold heat, icing should not be a problem.

LEAKAGE

Vacuum leaks in the intake system are most frequently encountered after a custom manifold installation. This doesn't preclude the possibility of a leak resulting from a carburetor swap. Fortunately, most carburetor-induced vacuum leaks are of sufficient proportion that there can be no doubt as to their presence. An unplugged vacuum pickup tube is easily located by the highly audible hiss caused by air being drawn into the opening. However, if a ported-vacuum hole is left open to the atmosphere there will be no indication of a leak at idle; the additional air will be drawn in only at higher RPM levels. ("Ported-vacuum" is not the same as normal "manifold vacuum;" at idle a ported-vacuum source is non-functional, but as engine speed increases the ported-vacuum source will become operational — that is, it will supply vacuum — at some pre-calibrated point in the engine speed range. Such sources are commonly used on late-model emissions engines.)

Double-pumpers and vacuum-secondary models with 50cc accelerator pumps must be checked carefully after installation to insure against vacuum leaks. When installed on certain intake manifolds, the accelerator pump housing or the pump housing attachment screws may contact the carb mounting flange, preventing proper seating of the throttle body against the manifold gasket. Some choke mechanisms are also incompatible with certain specialty manifold designs. Whenever a new carburetor or carburetor-manifold combination is installed for the first time, the carburetor mounting position must be checked carefully to insure that the carb is fully seated against the manifold gasket. If the it doesn't obviously seat squarely on the manifold, a portion of the carb is creating interference. Usually an outer edge of the manifold flange can be filed or machined to gain adequate clearance. A spacer between the manifold and carburetor base may be required.

Another cause of vacuum leakage is over-tightening of the carburetor attachment nuts or bolts. Over-tightening is the most common cause of a cracked throttle body; frequently the crack intersects a vacuum passage. Carburetor attachment hardware should be just snug enough to prevent engine vibration from causing the fasteners to loosen.

Super Tuning and Modifying HOLLEY CARBURETORS
Modifications for Street Performance

Ever tried to find a race engine fitted with carburetors that are even remotely related to the ones used on street vehicles? With the exception of the carbs used in classes where modifications are not allowed, you can't. The prevalence of such extensive alterations, and the advertising campaigns responsible for them, might lead you to believe that similar revamping is necessary if optimum performance is to be extracted from a high-performance street engine. No matter how radical a street engine is, the simple fact is that such a powerplant must function in an environment that bears little resemblance to a race track.

Even the most avid street enthusiast must make concessions in the name of driveability or family harmony. If a car hesitates, stumbles, bumbles, and dies at every stoplight, or is excessively sluggish at lower RPM, expletives that would make a politician blush will spew forth from the mouth of the occupant in the passenger seat. In truth, a carburetor used in street applications has a tougher job than its race-only brethren. On an oval track, drag strip, or road course an engine is either idling or at full throttle, but in street and highway use, engines must idle, cruise, accelerate at part and full throttle, and operate in several modes of deceleration.

These are the conditions for which Holley carburetors are designed, so the achievement of maximum performance a matter of choosing the right carb at the outset and tailoring it to a given engine/chassis combination. In fact, very few of the modifications discussed in Chapter 9 are applicable to street carburetors.

The modular construction of Holley four-barrel carburetors makes them inherently easy to modify. That doesn't imply that a component should be altered simply because it can be removed, reworked, and replaced easily. During the 40-odd years that Holley modular four-barrels have been in existence, virtually every part number has been tweaked to deliver optimum performance. In most cases, a carburetor is "spot on" for the application for which it was intended. If there are anomalies, they arise either from the myriad of installations that one part number must serve, or from the ability of many people to choose precisely the wrong carburetor for any given application.

In a sense, any aftermarket carburetor (whether direct replacement or high-performance) is similar to a rack suit. A single size may be worn by a wide variety of bodies, but if the suit is to fit perfectly, it must be taken in, let out, tapered, or flared to encompass the individual physique. So too, a carburetor may be tailored to accommodate the idiosyncrasies of a particular engine or driver.

MILLING OF FLAT SURFACES

When Holley builds a carburetor it assemble a group of repetitively pretested subassemblies. Each component is tested individually or in combination with other interrelated components and the carburetor is checked as a complete unit. As with any mass-produced item, the possibility of a problem with a brand new carburetor exists, although the probability is fairly remote. Some alleged carburetor experts have popularized the notion that warpage of the main body and metering block mating surfaces is a common occurrence. In a new carburetor, the possibility of excessive warpage is equal to the chances of finding an honest politician. With the passage of time, the repetitive heating and cooling cycles to which a carburetor is subjected may cause the main body and metering blocks to warp slightly, but being fairly thick and highly compressible, the metering block gaskets will seal effectively if flat surface distortion is less than .010-inch. This can be easily verified with a straight edge, and corrected with a medium-fine flat file, on the main body only. Under no circumstances should the metering block be filed. The surface that mates to the main body contains a number of ridges and two locating pins. Removal of the interior ridges, known as sealing beads, will impair gasket seal, and elimination of the locating pins will complicate gasket installation. The large ridge which runs around the outer edge of the metering block serves to limit gasket compression. Elimination of this ridge makes over-tightening of the fuel bowl screws possible and the resulting over-compression of the gaskets can cause the gasket material to block passages in the fuel block, thereby impeding or preventing fuel flow.

Holley metering block and float bowl gaskets, or those of similar quality, should be used in all cases, since the quality of many off-brand items is frequently inferior. If you've never disassembled a Holley four-barrel, you're in for a treat. The float bowl and metering block gaskets are treated with an adhesive that sticks like its life depends on it. The only reasonably efficient way to remove gasket remains is with spray-on gasket solvent, a healthy amount of elbow grease, and a gasket scraper.

FUEL BOWL VENT BAFFLING

While you've got the fuel bowls off, you may want to consider modifying the vent baffling. This is necessary only if your driving style consistently places your vehicle in "unusual attitudes." It is possible for fuel to spill out of the vent tubes if you roll, pitch, or yaw excessively. An easy fix is to install fuel bowl vent screen 26-39 or vent baffles 26-40 (2 inches long) or 26-89 (1-7/16 inches long). The screen breaks up fuel splash, but must be inspected periodically to insure that dirt and dust have not plugged the screen holes.

Designed primarily for use in the center pivot bowl, the vent baffle (or

No other carburetor provides the overall tuning flexibility of the Holley 4150/4160 modular design. Fuel bowls, main jets, power valves, secondary opening, accelerator pump volume/rate of delivery, and inlet needle-and-seat all can be altered easily to suit any specific application.

"whistle") extends the bowl vent path over and above the fuel. The whistle is ideally suited for severe acceleration where fuel is pushed back against the metering block surface of the primary fuel bowl. Since the baffle effectively closes the vent hole to fuel, there can be no spillover into the primary throttle bores.

A whistle is unnecessary in the secondary fuel bowl because, under acceleration, fuel is pushed back away from the vent opening. Nonetheless, if you routinely find your passengers bracing themselves against the dashboard during normal stops, your driving style may be such that fuel is being forced out of the secondary vent, into the sec-

Perhaps the most famous 2x4 factory induction system was this cross-ram setup for the early Z-28 Camaro. Intended for high-RPM racing use, this manifold with double-pumper Holleys also has made appearances on a number of street machines.

Warpage of the main body may occur on carburetors that have led a long, hard life. This body has been in service since 1969 and is still absolutely straight and flat. Don't consider machining the main body unless measurements indicate that the body has, in fact, warped.

Metering blocks are usually found in excellent shape, but if one has warped, milling won't solve the problem. The ridge that runs around the outer edge of the block serves to limit gasket compression when the fuel bowl screws are tightened. It, and other sealing ridges on the block surface, should always be left intact. If necessary, replacement fuel blocks can be obtained from Holley.

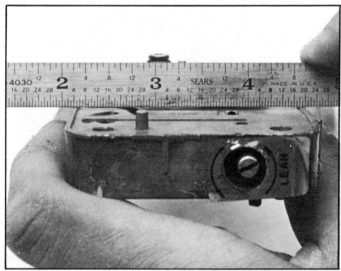

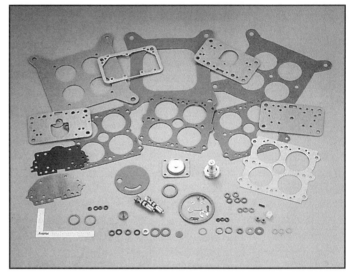

ondary throttle bores. Such sudden enrichment can flood the engine, causing it to stall, so if planning ahead for red lights isn't your strong suit, a secondary vent baffle may be in order.

OPTIONAL FUEL BOWLS

For the majority of street applications, a 600cfm Model 4160 with vacuum secondaries, or a 650cfm Model 4175 is an excellent selection, as they both supply adequate air flow capacity and are reasonably priced. However, these carburetors are fitted with standard side-pivot float bowls which do not provide for optimum fuel flow control during hard cornering. For aspiring young road racers, salvation is available in the form of Fuel Bowl Conversion Kit no. 34-2 for 4150/4160 series carbs and 34-4 or 34-5 for model 4165s. (Conversion kits are not offered for 4175s.) These kits contain two center inlet "dual feed" bowls and offer an installation advantage in that fuel lines may be connected to either side. A center pivot float design is typically less sensitive to changes in side-to-side vehicle movement (cornering) and therefore offers improved fuel level control. The inconvenience of repositioning the O-rings on the fuel transfer tube is also eliminated. Also available is part number 24-15 for model 2305 two-barrels.

Even more inconvenience can be eliminated with Holley's Quick Change Jet Kits, which include special float bowls that allow jets to be changed without removing the float bowls. The bowls in Quick Change kits contain two screw-in plugs which, when removed, allow direct access to the jets. A special jet-changing tool that holds the jets is included, as are plugs and gaskets. Part number 34-24 is designed for Model 4500 carbs and includes both primary and secondary bowls; part

The modular design of Holley 4150/4160 carburetors has translated into many iterations of individual components and subassemblies over the years. Most rebuild kits contain several gaskets that are seemingly the same. They're not. It is absolutely essential that new gaskets are matched to the ones they're replacing during tuning or rebuilding procedures.

number 34-25 contains only a primary bowl for Model 4150/4160 carbs. Part number 34-26 contains only a secondary bowl for Model 4150s with secondary accelerator pumps; part number 34-27 includes a bowl for Model 4150 carbs with vacuum secondaries. (The jet removal tool, part number 26-68, is not included in the secondary float bowl kits.)

When periodic rejetting is a consideration, another desirable modification to 4160 carburetors is the installation of a secondary metering block (containing removable jets). All Model 4150 carburetors are so equipped, but Model 4160s will have to be modified with Kit 34-6. Essentially, this kit consists of the "jet-table" metering block, which may be installed in place of the original metering plate that contains fixed metering orifices. Gaskets, longer screws, and a longer primary-to-secondary bowl fuel transfer tube are also supplied. Appropriately sized jets must be purchased separately.

For street vehicles that venture to a drag strip only on rare occasions, the conversion kit is probably not required, since there will be little need to alter secondary jetting. But, for regular competition or "heavy" street use, the ability to replace secondary jets is highly desirable.

In some instances, drilling the secondary main metering restrictions of the metering plate is a viable alternative to the 34-6 kit. This approach assumes that secondary fuel metering is excessively lean. The holes in the bottom of the plate may be enlarged by a few drill sizes if necessary. Drilling should be done by hand, with the bit held in a pin vise and the plate should be blown free of chips with compressed air prior to reassembly. If the thought of drilling the plate is unnerving, purchase a different one. The chart in the Appendix lists restriction sizes of the various part numbers of Holley metering plates. When a replacement plate is selected, the idle hole diameters should be the same (or as close as possible) as those in the original plate. Both the 4160 and 4175 use the plates listed under part number 134-X or 134-XX; the single or double digit suffix is much the same as a jet number. In most but not all cases, the suffix also matches the number stamped on the plate.

SECONDARY ACTUATION

Whenever there is concern about the metering of fuel to the secondary throttle bores, interest in the rapidity with which the rear barrels are activated is piqued. Secondary vacuum diaphragm spring kit no. 20-13 contains a variety of springs, each of which provides a different opening rate. Holley four-barrels fitted with vacuum operated secondaries rely on airflow to open the secondary throttle plates. One of the most appealing aspects of this system is extremely smooth operation ranging all the way from idle to maximum RPM. Since the engine receives only as much airflow capacity as it requires, power output can be optimized all along the RPM curve (as compared to an equivalent carburetor with mechanical secondaries). Vacuum secondaries may not look as exotic as race-type double-pumpers, but when the diaphragm spring is properly calibrated, they are superior in the "real world" of street driving.

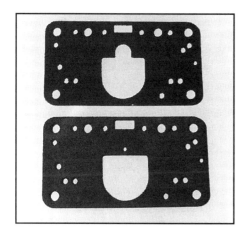

Holley offers different styles of metering block gaskets, so it is advisable to carefully match the new gasket with the one it will replace. The top gasket is intended for blocks with an accelerator pump transfer tube. If it is used on a carburetor not so equipped, fuel from the pump circuit will be drawn into the engine by manifold vacuum and transmitted through the power valve chamber.

For normal operation, the standard fuel bowl vent baffle is entirely adequate. However, to prevent fuel spillover during hard cornering, acceleration, or braking, a special vent extension, often called a "whistle," may be required.

For road-racing or gymkhana competition, center inlet float bowls provide additional fuel control benefits. It is easy to convert from side inlet to center inlet (dual feed) bowls with Holley kit number 34-2 (for 4160) or kit 34-4 (for 4165 carbs).

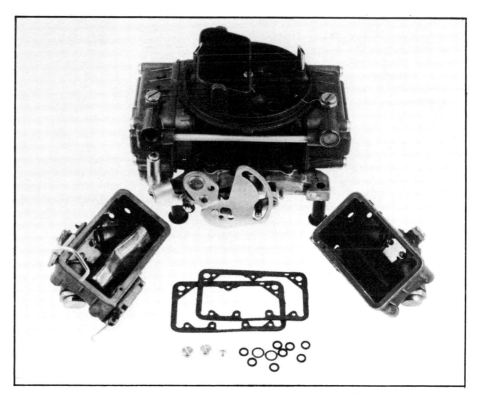

The difference between a double-pumper and a vacuum secondary four-barrel easily may be seen in this photo. Note great similarity between float bowls on the secondary side. The basic casting is identical, but since no accelerator pump circuit is used with vacuum actuation (left), the pump hardware is not installed and the fuel supply passages are not drilled.

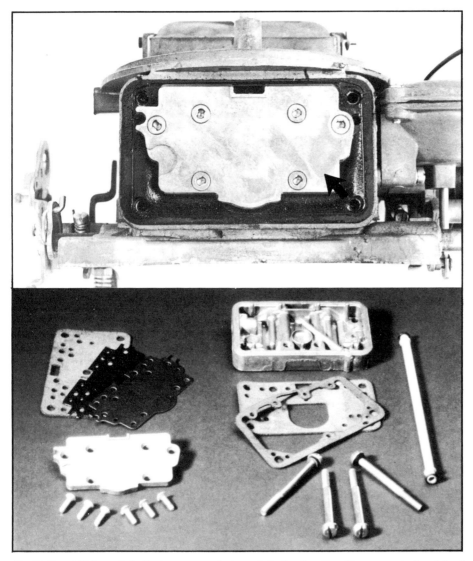

On Holley 4160 models that use a metering plate (top) rather than a metering block, the fuel metering orifices and delivery passages are incorporated in the metering plate. If such a carb will be used in a racing application where main metering changes are frequently required, it is possible to convert to a 4150-style metering block (with replaceable jets) with kit number 34-6. Shown at right (bottom photo) are the parts that replace the metering plate (left). Note, also, the metering block assembly is sandwiched between the main body and the float bowl, making the carb longer overall.

Occasionally, the secondary spring in a carburetor designed to fit a variety of engines may be too stiff to allow the secondaries to open quickly or fully. Changing the spring within the secondary diaphragm housing is a quick, easy remedy. Secondary spring kit 20-13 contains seven springs, each with a different relative rate (stiffness). Lightening the spring rate causes the secondaries to open sooner, while a heavier spring will delay opening. Only careful trial-and-error testing over a measured acceleration course (drag strip) will determine which spring provides the optimum opening time. The ideal situation is to use the lightest spring possible that will allow secondary actuation without causing an engine to stumble.

ACCELERATOR PUMP

The secondary throttle opening rate is not the only aspect of carburetor operation to alter the tendency of an engine to stumble upon full throttle acceleration. The accelerator pump circuit also comes into play. As noted previously, Holley offers a variety of pump cams which allow both the volume and duration of pump discharge to be varied. Depending on the cam used, full lift may not be reached until comparatively late in the throttle opening cycle. Such "long" cams may compensate for an excessively quick secondary opening rate. If a sufficient volume of fuel is discharged as the carburetor is brought to wide open throttle, the engine may never feel the lean condition that occurs when the secondaries open too quickly.

Obviously, the most consistent performance will result from a properly selected secondary opening spring. But there are situations in which the proper spring has a tendency to cause a stumble when an engine is brought to wide open throttle under "unique" operating conditions. Changing the pump cam, or varying the mounting position (most cams have two mounting holes, each of which provides a slightly different pump cycle), can solve such problems.

For street engines equipped with a high overlap camshaft and a single-plane manifold, an off-idle stumble during normal acceleration can be a problem. Changing pump cams or enlarging the discharge nozzle diameter can compensate somewhat for poor off-idle performance, but the problem should not be considered truly solved simply because a degree of driveability has returned. In actuality, low manifold vacuum created by the cam/manifold combination is providing the main discharge nozzle with a relatively weak signal. This delays activation of fuel flow through the discharge nozzle, hence the stumble. Discharging a greater volume of fuel through the accelerator pump circuit simply "covers up" the momentary leanness that occurs during transition from the idle to the main fuel metering circuits. Although the accelerator pump "cover up" is an easily executed fix, it may not prove satisfactory under all operating conditions. In instances where the throttle is not returned fully to the idle position, as during quick, repeated deceleration-to-acceleration transitions, the pump may not completely refill with fuel. The reduced volume of available fuel in the pump may be insufficient to prevent a stumble or sputter.

POWER VALVES

The power valve and its effect on a high-performance street engine constitutes a vast gray area for many enthusiastic carburetor tuners. Although the rhetoric produced by some pseudo experts promises that a power valve will do for a sluggish engine what Dr. Love's Elixir was purported to do for a sluggish body, the fact of the matter is that improvements in performance are minimal in both cases.

Changing to a power valve with a higher opening point (as in stepping from a 6.5 to an 8.5) will bring an engine to maximum richness more quickly, but that's about the only effect. Depending upon the appropriateness of the carburetor for the specific application, the effect of changing power valves may vary from none whatsoever to distinctly noticeable. However, any change in performance will exist only during the few moments following a sharp stab on the accelerator pedal. Once a valve is open, the amount of enrichment is virtually the same, regardless of the manifold vacuum figure at which the valve operates.

Another approach to fuel mixture optimization involves use of a two-stage power valve. The progressive opening system allows leaner jetting than might otherwise be practical because the first stage will open during moderate acceleration. Partial enrichment at relatively high manifold vacuum levels serves to minimize or eliminate the lean misfire situation that is so characteristic of carbureted "emissions" engines during acceleration.

Two-stage power valves were developed specifically for relatively heavy recreational and utility vehicles that require some enrichment during acceleration at medium throttle. In theory, such a scheme is also suitable for a high performance engine. But two-stage power valves can create excessively lean conditions because they have less flow capacity than compared to a standard single-stage valve. It is specifically for this reason that two-stage power valves should never be used in a carburetor that sits atop a high performance engine with a high fuel demand.

STREET JETTING

As with most things automotive, carburetor jetting is formulated through compromise. At one end of the requirement spectrum, fuel economy necessitates a degree of leanness; at the other, the production of horsepower dictates a

The incredible flexibility of the Holley Model 4150/4160 design makes it possible to adapt them to any imaginable application. Here a pair of Holleys feed a GMC-blown small-block Chevy. The unique Holley accelerator pump design allows special "tailoring" for immediate throttle response, even with superchargers. Note the high-volume Reo pump mounted on the forward carb.

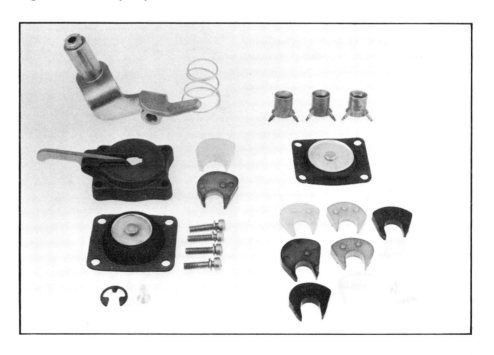

Holley accelerator pump cams and delivery nozzles can easily be altered for virtually any engine with simple replacement component (at right). For engines needing a healthy dose of fuel when the throttles are cracked wide open, a large-volume (50cc) accelerator pump assembly (left) can easily be installed in place of a stock 25cc version.

Super Tuning and Modifying Holley Carburetors

The key to successful street performance is to stay away from "exotic" tricks. Select the proper carb for your specific applications and you probably won't have to modify it at all. Fine tuning of main jets and secondary activation points are usually all that's needed.

considerably richer air/fuel mixture. In spite of stoichiometrically correct mixtures, an internal combustion engine must function within a vast arena of ever-changing variables. Proper jetting merely offers the best compromise for the widest number of variables.

Where one driver will sacrifice some amount of power in the name of improved fuel economy, another will find such a tradeoff totally unpalatable. Since it is impossible to universally define "optimum operation," determination of jet sizes that may be applied "across the board" is equally impossible. As supplied, most Holleys are jetted to offer the best compromise for a variety of specific applications. Careful experimentation based on individual preference is the only method available to achieve that elusive ultimate compromise.

DOUBLE-PUMPERS FOR THE STREET

The 4700 series double-pumper four-barrels (part nos. 0-4776 through 0-4781) were released as "after-market" units for a variety of performance applications. However, these carburetors were designed to function primarily on race cars. Consequently, a great deal of rework is necessary for unconventional "street" use.

Anticipating the fuel delivery requirements of racing engines, Holley engineers specified comparatively rich jetting for these carbs. A 600cfm 0-4776 double-pumper is jetted five steps richer than a 600cfm 0-6619 (which is classified as an "emissions design street performance" carb). Both carburetors possess the same 1-1/4 inch primary venturi diameter and 1-9/16 inch throttle bore diameter. Obviously, the 4776 would deliver significantly reduced fuel economy under cruise conditions. What is not so obvious are the differences in air/fuel ratios at idle and during wide open throttle.

A great deal of insight is not required to realize that the double-pumper may be easily rejetted to provide an air/fuel ratio that is identical to that of the 0-6619. However, were this to be done, an excessively lean air/fuel ratio would result during full throttle operation because the power valve channel restrictions in many double-pumpers are relatively small. (They were sized to function in concert with the relatively large original jets.) In order to achieve a proper air/fuel ratio at full power, the PVCR's would have to be enlarged to compensate for any reductions made in jet size.

The important point to remember when altering air/fuel mixtures through a combination of PVCR and jet changes, is that total flow area, as opposed to orifice diameter, must be considered. Assume that the carburetor to be modified by a street-racer-turned-economy-driver is fitted with number 69 jets and .040-inch PVCR's. Using the basic formula for computing area of the circle, $A = \pi R^2$ (Area equals Pi times the square of the radius of the circle), total flow area can be determined as follows:

Main jets play a big role in determining the overall fuel metering efficiency of a carburetor. Another important part of the air/fuel metering equation is the power valve. With highly modified engines, it is often necessary to experiment with both jet size and Power Valve Channel Restriction (PVCR) diameter to develop maximum power at wide open throttle along with acceptable part-throttle drivability.

Jet size area

$A = \pi \times .035^2$, $(\pi = 3.14159)$*
$A = \pi \times .001225$
$A = .0038$ sq. inch (area of a #69 jet)

*Note: According to the Holley jet chart a #69 jet has a .070-inch drilled hole (with a radius of .035-inch). Computation is for area of a single jet.

PVCR area

$A = \pi \times .020^2$
$A = \pi \times .0004$
$A = .00126$ square inch.

Adding the results of both equations (.0038 + .00126) provides a figure of .00506 as the total flow area (jet + PVCR) per throttle bore. Assuming that jet size is to be reduced by five steps, to a #64, the revised flow area figures would be:

Jet area = .00321

Total area (Jet + PVCR) should equal .00506

If the total area is to remain the same, subtracting the new jet area from the original total area will provide a "corrected" PVCR area. In this instance:

PVCR area should be .00185 (.00506 - .00321) (total area less jet area)

Working the area formula from a different angle the radius and diameter of the corrected PVCR may be determined:

Area = $\pi \times$ radius^2
$.00185 = 3.14159 \times r^2$
$\dfrac{.00185}{3.14159} = r^2$

$.00058 = r^2$
$.024 = r$

diameter = radius x 2
diameter = .048.

Therefore, when jet size is reduced from .070-inch to .064-inch, PVCR size must be increased .008-inch, from .040-inch to .048-inch. Since area is the consideration in question, it should be noted that increases or decreases in PVCR diameter will be influenced by overall jet size as well as the degree of change. Note that the PVCR had to be enlarged by .008-inch to compensate for a decrease of .006-inch in jet size. This is due to the area differences between a .040-inch and .070-inch orifice.

Another area where a double-pumper demonstrates rich mixture characteristics is in the idle circuit. Typically, the diameter of the idle feed restriction can be reduced .002-inch to .003-inch, which will lean the idle mixture without affecting the off-idle performance. Unfortunately, locating replacement idle feed restrictions is difficult unless you have access to the Holley manufacturing facility (although a few companies do offer brass restrictions that may be drilled to the desired size). An alternative approach is to enlarge the idle air bleed .001-inch to .002-inch. A cautious hand is required in performing such operations. Once the bleed is made too large, you face the same problem that made drilling the bleed necessary in the first place. Modern science has not yet found a way to drill a hole that is smaller than the one that already exists. Another word of caution: be sure you drill the idle and not the high speed bleed.

If you do get carried away with drilling the bleed, it is possible to enlarge the idle feed restriction to compensate for the error. However, these two orifices have a distinct size relationship to each other and it should be maintained. It should also be noted that in some carburetors the idle feed restrictions are not easily accessible.

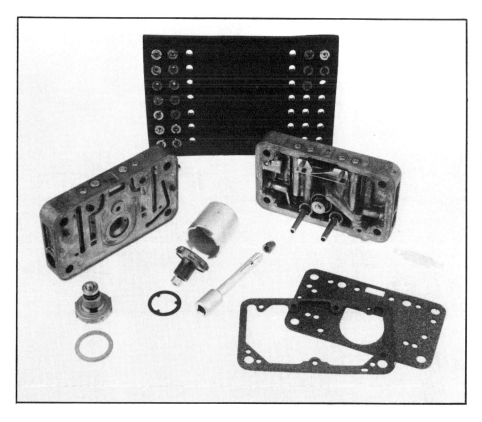

Easy access to the main jets and power valve makes tuning a Holley carburetor relatively easy. Although a 1-inch wrench for the power valve and a screwdriver are the only tools required, special wrenches, which reduce the potential for damage, are available.

Every street-driven vehicle should be equipped with a good air cleaner. This low-restriction model can actually increase airflow because it smoothes the entry path for incoming air. Some dyno tests have shown a 10-horsepower increase compared to runs made without an air cleaner.

Many carburetor modification companies offer screw-in air bleeds which may be easily changed. If an air bleed has been damaged or drilled to too large a diameter, it may be possible to save the carburetor by drilling and tapping the bleed hole to accept a screw-in bleed.

Super Tuning and Modifying Holley Carburetors **65**

Super Tuning and Modifying HOLLEY CARBURETORS
Turbo/Supercharging Modifications

Since the late Sixties, Holley carburetors have been considered the essential four-barrel for every type of motorized vehicle, from high-performance street machines to ski boats to competition-only race cars. Most racers and performance enthusiasts favor Holley four-barrel designs because they offer excellent performance "out-of-the-box," and can be easily modified for virtually any highly specialized application. Yet the aura of mystery that surrounds carburetors in general can become a stumbling block when a carburetor is to be used on a supercharged or turbocharged engine.

In truth, modifying a Holley for fuel delivery to a supercharged or turbocharged engine is only slightly more difficult than changing jets or power valves. Most blower/turbo kit manufacturers design their products to function effectively with the carburetor already on the engine. This practice helps to minimize the expense of a turbo installation, but creates a potential problem in that replacement parts such as floats, jets, and power valves are not readily available for many original equipment carburetors. With the widespread popularity enjoyed by the Holley 4150/4160/4165/4175 series, many small parts are available at performance-oriented auto, marine, or RV shops, so everything needed to custom build a "blower/turbo Holley" may be easily located. Obviously, an understanding of the relationship between the carburetor and a blower/turbo induction system is helpful in determining areas to be considered for modification.

Neophyte blower/turbo enthusiasts should be aware of the two basic approaches to supercharging and turbocharging: Blow-through and draw-through. These terms refer to the position of the blower/turbo unit with respect to the carburetor. Blow-through installations typically maintain the carburetor in the original location and pressurized air from the blower/turbo outlet is ducted to the carb inlet. This arrangement is similar to the one used by Ford on the Paxton blower-equipped 312 engine. Draw-through configurations require movement of the carburetor to a remote location in order to allow the blower/turbo to draw air and fuel from the base of the carburetor and deliver the pressurized air/fuel mixture to the intake manifold. In terms of conventional supercharging, a draw-through installation may be equated to a GMC Roots-type blower.

Each method has distinct advantages and disadvantages. The blow-through method pressurizes the entire carburetor and usually requires that certain steps be taken to prevent fuel from being blown out of openings (vents and throttle-shaft opening) that are exposed to atmospheric pressure. On the plus side of the blow-through balance sheet, the carburetor may be maintained in the standard location, allowing retention of the original choke and manifold heat to assist in cold starting and operation.

Conversely, draw-through installations require relocation of the carburetor, a custom choke, and manifold heat mechanisms. But, since the carburetor is not pressurized, it does not require sealing. Nor does the fuel pump have to be modified. This is the major reason that the draw-through approach has become more popular than the blow-through.

DRAW-THROUGH MODIFICATIONS

No matter what the case, fuel metering in a blower/turbo application presents some difficult problems. In a draw-through system it is possible for the engine to be in a fully loaded condition, yet the blower/turbo is still pulling 5-6 inches of vacuum across the carb. If the power valve activation point is below this vacuum level, the valve will close, cutting off 30-40% of the overall fuel delivery and the engine will "lean destruct" quickly. To solve this, the power valve activation point must be kept relatively high (8-10 inches) or the valve must be externally referenced.

External referencing is the best and most common way to ensure relatively normal power valve operation. That is, the power valve must sense "true" vacuum conditions in the intake system through an external line which runs from the carburetor power valve chamber directly to the intake manifold. This may be accomplished by first blocking the standard power valve vacuum pickup hole in the throttle body. It will also be necessary to plug the screw hole in the throttle body that's adjacent to the pickup hole if a screw is not installed in this location. A suitably sized piece of lead shot, pressed into the throttle body with a hammer and punch, is adequate. JB Weld can also be used, as can a screw-in plug, if the hole is tapped. Regardless of the means chosen, it's imperative that the plug be securely installed. Use of Loctite on threaded plugs is highly recommended. Next, a new pickup hole must be drilled either in the side of the throttle body or in the main body.

With most carburetors, the easiest method of externally referencing the power valve is to use the passage that normally brings ported vacuum to the metering block. This can be accomplished by drilling a hole on an angle from the underside of the main body into the power valve vacuum chamber. After the new hole is drilled,

This 355ci small-block Chevrolet is equipped with two 650cfm Holley carbs and a 6-71 blower. It produced 626 horsepower with only minimal reworking of the carburetors.

the original path to the throttle body must be blocked. Once this is done, a small line can be run from below the turbo/blower to the former ported vacuum fitting in the metering block and *voila* — you have an externally referenced power valve.

If this configuration can't be used, a hole can be drilled in the throttle body. This hole need only be large enough to accommodate a small-diameter tube that will subsequently be pressed into place. If suitably sized tubing cannot be located, a Holley main jet extension (part no. 26-21) will work satisfactorily. A tee fitting inserted in a normal manifold vacuum line may then be used as a source for

The best way to eliminate pressure differential problems in a blow-through installation is to enclose the whole carburetor in a box. Since pressure is the same inside and outside the carburetor, there are no concerns with blowing fuel into the relatively low-pressure atmosphere.

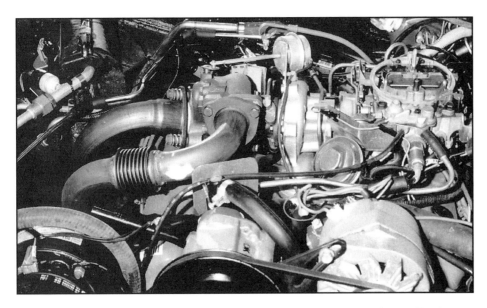

In 1980, Pontiac introduced a turbocharged V-8 equipped with a Rochester QuadraJet. Performance and "tune-ability" generally can be improved by replacing the stock carb with a 4175-type Holley.

manifold vacuum if all tapped holes in the manifold are in use. Irrespective of the method used, it's advisable to install a one-way valve in the line between the intake manifold and the carburetor to prevent power valve damage at high boost levels, or in the event the blower "sneezes." If this is done, it's also necessary to drill a small relief hole. Otherwise, vacuum will be maintained in the power valve chamber after the engine goes into boost.

If a carburetor is equipped with a secondary power valve, it too will have to be externally referenced. In addition, the secondary PVCR's should be enlarged slightly to enrich the full throttle mixture. The PVCR's are the smallest-diameter orifices in the power enrichment circuit and, as their name implies, they constitute the restrictions that control fuel flow. As a starting point, these holes should be drilled two index numbers larger. Care should be taken to prevent drilling entirely through the metering block. Enlarging the PVCR's can best be accomplished using a pin vise to hold a drill bit, or using a drill press or vertical mill with the metering block clamped in a proper holding fixture. It is important to note that allowing the bit to wobble while drilling will cut the hole oversize, possibly resulting in over-richness whenever the power valve opens.

If you are wary of drilling into the metering block, it is also possible to block off the secondary power valve circuit. A Holley power valve plug can be installed (part no. 26-36) to deactivate the circuit, but the secondary main metering jets will have to be enlarged to compensate. Typically, this will require 6-8 steps larger jets but will vary considerably, depending on the specific carb. The formula below will approximate the diameter of the new jet size, which must equal the old jet size area plus the PVCR area.

DIA new jet = DIA old jet + DIApvcr

A final modification may also involve installation of an accelerator pump that will provide a larger shot of fuel. Typically, a draw-through blower/turbo installation mounts the carburetor some distance from the original carb position on the intake manifold. Since the air/fuel mixture must now travel a greater distance in order to reach the combustion chamber, the accelerator pump must supply a greater volume of fuel to gain immediate response. The simplest solution, of course, is to install the large-volume Reo pump kit. There are only two cams for this pump, so the best way to tailor the delivery is to increase or decrease the size of the discharge nozzle until the engine responds crisply without over-enrichment.

BLOW-THROUGH MODIFICATIONS

Blow-through blower/turbo installations require other types of modifications. Since the carburetor will be pressurized, the standard hollow brass floats will be subjected to pressures that may collapse them. Substituting nitrophyl (hard, formed plastic) floats for the standard brass units alleviates this situation. Holley part number 116-3 is designed for center-pivot (dual-feed) type float

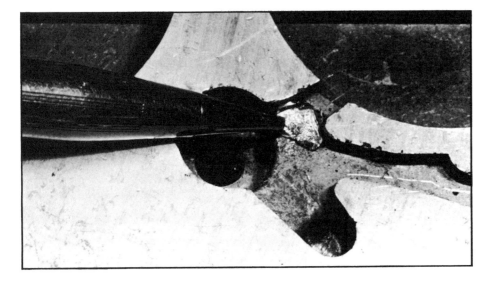

The first step toward externally referencing the power valve is to locate the standard vacuum source that leads to the power valve chamber. When viewing the throttle body from the underside, the vacuum delivery passage will be just to the side of the countersunk bolt hole (which may or may not be used, depending on the carb model). The pickup hole should be blocked with a piece of lead shot or JB Weld, or tapped for a screw-in plug.

bowls and part number 116-1 fits side pivot "flat" bowls.

If the entire carburetor is sealed in a pressure box, very little modification is required since internal and external pressure will be relatively equal. This is the approach that was taken with the Paxton centrifugal blower used on the 1957 Ford 312 engine. However, the Paxton bonnet is somewhat restrictive. Having been designed for a relatively low airflow application, it may be easily improved. In the 1980s, when turbocharging became popular, H.O. Racing Specialties, Lawnsdale, Cal., offered a system that pressurized only the top of the carburetor. The H.O. bonnet design may provide an insight into a more efficient method of enclosing the entire carburetor in a box.

When only the top portion of the carburetor (as opposed to the entire unit) is pressurized, a slightly different technique is usually needed. Depending on the carburetor model, it may be necessary to seal those openings (such as around the throttle shafts) that connect a high pressure area (inside the carb venturi) to one of atmospheric pressure. This is done to prevent fuel from being blown out of the carburetor. One of the simplest methods of sealing the throttle shafts is to create an air seal using blower/turbo pressure.

The air seal principle is quite simple. High pressure air will always move toward a low pressure area. In a blower/turbo environment, atmospheric pressure constitutes the low pressure area and the clearance between the throttle shafts and body provides a path to that lower pressure. By simply applying equal pressure to the shaft-to-body air gap, the pathway is effectively blocked, thereby preventing fuel loss.

Drilling small holes in the throttle body just below and on each end of, the throttle shaft will provide an entry path for the pressurized air. These holes should be approximately .060-inch in diameter and the drilling should be done on a drill press so that depth may be carefully controlled. After removal of the throttle plates and shafts, four holes should be drilled from the underside of the throttle body. Use care when drilling so that when the drill bit strikes through into the throttle shaft bore, the other side is not scratched.

The hole indicated (arrow) is normally a source for a ported manifold vacuum. Drilling an intersecting hole between it and the power valve chamber is the easiest way to externally reference the power valve. The intersecting hole can be drilled from the underside of the main body.

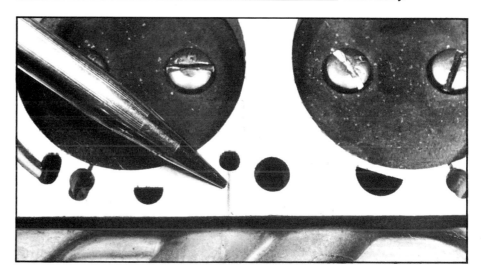

The vacuum source hole that connects to the power valve chamber is drilled completely through the throttle body, so an intersecting hole can be drilled from the side to provide an external vacuum source. It's a good idea to draw a line before drilling to make sure you won't be drilling into anything you shouldn't. With some carburetors, it may be necessary to drill in at an angle.

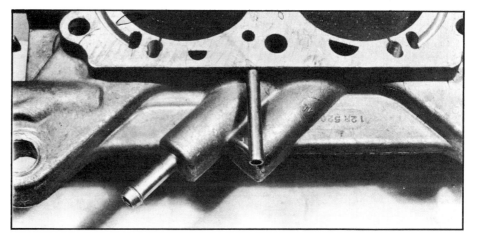

A tube may be pressed into place and a neoprene hose connected. The other end of the hose should then be attached to a manifold source on the pressure side of the turbo/blower.

Remove all burrs prior to reinstalling the throttle shafts.

Once the throttle body is drilled, a 1/4-inch to 1/2-inch spacer plate

Super Tuning and Modifying Holley Carburetors

These two holes constitute the restrictions (PVCR) within the power enrichment circuit. Drilling them slightly oversize is recommended for some turbo applications.

This metering block, viewed from the fuel bowl side, has a plug installed in the power valve hole. This is an alternative to externally referencing the secondary power valve (if the carburetor is so equipped); however, the secondary main circuit (jets) also will have to be enriched. The primary power valve should never be blocked.

Solid plastic (nitrophyl) floats prevent float collapse that can occur in blow-through installations.

It's hard to beat the appearance and power of two Holleys sitting on top of a "Jimmy" blower. Mounting the carbs sideways allows use of dual-feed float bowls.

should be drilled to match and cross-drilled to intersect the vertical holes (see accompanying illustration). Tubes pressed into the cross-drilled holes will provide a convenient means for attaching pressure hoses from the turbocharger. (The cross-drilled holes may be tapped for threaded fittings, if preferred.) Naturally, the carburetor gasket should also contain properly located holes.

Virtually every blow-through blower/turbo kit manufacturer specifies the need for throttle shaft sealing when a Quadra-Jet is used. It appears, however, that the tolerances on a Holley may be considerably tighter and in certain applications shaft sealing may not be necessary. Some installation specialists have designed blow-through systems which have not required sealing of the throttle shafts.

When using a carburetor with vacuum secondaries, the underside of the secondary diaphragm must be sealed. Some kind of boot should be installed over the secondary opening rod and this chamber connected to boost pressure. In this way the secondary diaphragm will be referenced to the same relative pressure as the airhorn and the secondaries should function normally. However, use of a carburetor with mechanical secondaries is much more practical since it alleviates the need to seal the diaphragm housing.

Even with these modifications the average blow-through system will be limited to relatively modest boost (under 10 pounds) because the carburetor (any carburetor!) is simply not able to cope with high-density air delivered to the carb inlet by a high-boost blow-through system. All standard carbs are designed with the understanding that atmospheric pressure will deliver air of approximately average density into the main venturi. Fuel circuits are subsequently designed to deliver the correct proportion of fuel into this air for proper combustion. However, under full or partial boost, the air delivered by a blow-through system to the carb may be twice or even three times as dense as "normal" air. The carb has no way of telling that the incoming air is "denser"

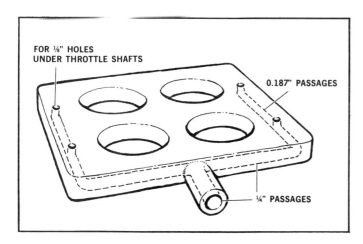

When high-pressure air is applied to the carburetor throttle bores of a blow-through turbo system, some method should be devised to keep air/fuel from being blown out of the throttle shaft bores in the throttle body. One of the most effective means is to use turbo boost pressure to create an air seal around the shafts. A drilled plate is placed under the carb and matching holes are drilled through the throttle body to the shaft bores. Boost pressure from a hose connected to the turbo manifold enters the plate and is blown upward through the connecting passages to the throttle shaft bores.

This custom Corvette engine is equipped with dual turbos and a blow-through system. The Holley carb has been modified as described in text.

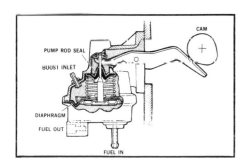

In blow-through systems using a mechanical fuel pump, the upper chamber must be referenced to boost pressure so the pump will develop sufficient pressure to deliver fuel to the carb float bowls (which are also pressurized by boost). A sealed operating rod is also required so boost pressure does not escape into the crankcase.

This plumber's nightmare is actually a Holley carb (mounted on an intake manifold) designed for a blow-through turbo installation. The carb features an air-sealed throttle shaft, boost-referenced vacuum-secondary chamber, and boost-referenced fuel pump.

and requires considerably more fuel for proper combustion. (Remember, carb functioning is based on pressure differential and air "velocity" through the venturi.)

It is possible to counter this problem somewhat by "over-enriching" the secondary circuit of the carb. But this is no real solution. A "rich" secondary will compensate somewhat for the dense boosted air, while still retaining some semblance of proper fuel delivery during non-boost/primary operation. But the maximum boost pressure must be kept below the point at which secondary fuel flow falls short of the "density" increase. (The only real solution for boost pressures of about two atmospheres or more is a third delivery circuit, but this involves re-engineering the entire carburetor.) As a practical matter, the only currently available solution is to restrict maximum boost pressure and carefully adjust secondary fuel delivery to compensate.

In all cases, on blow-through installations, fuel pressure must be 6 psi above the boost pressure. This will insure that the carburetor has an adequate supply of fuel at all times. It can be done with an electric pump and regulator by referencing the regulator to boost pressure instead of atmospheric. Care should be taken to insure that the electric pump selected can provide 6 psi pressure above the maximum boost pressure. If a mechanical pump is retained, the area over the pump diaphragm must be sealed off from the crankcase. By connecting this chamber to boost pressure with an external line, the pump will always be supplying fuel at its normal pressure above boost pressure.

While some of these modifications may sound intricate, they are relatively easy to perform. Competent help may be obtained from many supercharging and turbocharging specialists throughout the country.

Super Tuning and Modifying Holley Carburetors **71**

Super Tuning and Modifying HOLLEY CARBURETORS
Modification for Competition

Holley carburetors have long dominated all forms of racing. In addition to oval track racing success, the unique Holley double-pumper is virtually the only carb used by most drag racers.

With the highly competitive nature of oval track, Pro Stock, Competition, and Modified Eliminator drag racing, it is understandable that extraordinary time and dollar expenditures are put forth for the express intent of improving race track clockings by few hundredths of a second. One area in which racers spend excessive amounts of money is carburetion. In the late 1960s, a limited number of specialty shops began modifying Holley carburetors for specific racing applications. As use of highly modified carburetors spread, the population of modification specialists increased, and the prices they demanded began to rise significantly.

Much of the cost increase was based on the simple fact that it is difficult to find a racer who isn't firmly convinced that extensively modified carburetors are an essential ingredient in any race-winning effort. A look at the record book may indicate that a little carb magic is indeed necessary for admittance to the "Winner's Circle." Unfortunately, many racers are putting their cash on the line for magic, only to become the victims of a sleight-of-hand act. To put it bluntly, many alleged carburetor magicians are really ripoff artists.

Those are strong words but the evidence compiled and evaluated by the most knowledgeable people in the business — the Holley Carburetor engineering department — is conclusive. Many people modify carburetors for racing but most don't really understand what they're doing, nor do they have the proper equipment for testing the carbs after modification.

Holley engineers are not unilaterally opposed to modifications. What they do resent is the fact that racers are being charged hundreds of dollars per carburetor for modifications that are not necessary; most of the desired performance improvements could be accomplished by the racer himself with minimal time and expense. Many super exotic modifications and calibrations are largely cosmetic and offer no potential for performance improvements.

Under normal circumstances a large corporation doesn't care what consumers do to company products once they have purchased them. However, some time ago, Holley field technicians began encountering a growing number of racers who were complaining about their "carburetor problems." When Holley reps investigated, they found, that the carbs had been "modified" beyond a repairable state. But the racer wasn't prepared for that type of answer. After all, it was a Holley carb (when he bought it, at least), so the Holley rep should be able to fix whatever was wrong.

After reviewing the situation, Holley purchased carburetors from a number of well-known modification companies and tested them on race cars and at their engineering headquarters. The information gathered by the Holley engineers is enlightening. One supposedly "trick" carburetor leaked fuel out of the high-speed air bleeds and had to be rebuilt before it could even be operated in a flow test. Another carb had an idle circuit installed on the secondary side. Unfortunately, the genius who performed this operation neglected to machine a discharge circuit in the main body and throttle body, so the entire circuit was useless.

Before attempting to perform modifications, or before having them done by someone else, it is supremely helpful to understand a logical reason for each modification, its application, the possible benefits, and potential detriments. It will also be supremely helpful to your bank account to deal with a reputable modification company with a proven track record.

Another option is to purchase a carburetor that has already been modified for competition use — by Holley. The Pro-Series HP carburetors are designed to be taken out of the box and bolted directly onto a race engine. Available with air flow ratings ranging from 390 to 1,000 cfm, carburetors in the Pro-Series are available with either vacuum or mechanical secondaries. Common features include matched venturis with specially contoured entries, no choke tower, down-leg, double-step booster venturis, (for a stronger vacuum signal), machined throttle body, metering block mating surfaces, high flow metering blocks, and needle-and-seat assemblies.

Certainly some tweaking of a Pro-Series carburetor may improve performance levels, but it's also quite possible that not even a jet change will be required to produce maximum horsepower. Even if additional modifications are required for a strange and unusual engine/chassis combination, a Pro-Series carburetor provides a better starting point.

The Ford SOHC 427 was one of the many performance engines to be fitted with Holley four-barrels. The Model 4160s gave terrific performance with total reliability as they were part of a well-engineered induction system. Note the connection between the vacuum secondary diaphragm housings.

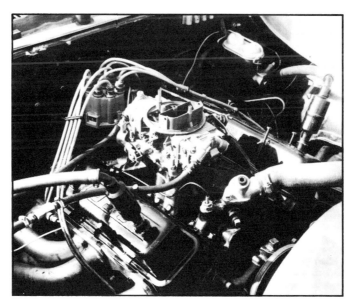

Some classes restrict carburetor modifications to basic internal parts replacement. This 750 cfm carb, installed on a 331ci Super Modified Camaro, had only the jets and needle-and-seat assemblies changed, yet it can run over 129 mph in the quarter mile.

CHOKE SHROUD (HOUSING) REMOVAL
(All Applications)

This is undoubtedly the most popular mod. It is also easily visible, so the competition may be intimidated because you have a "super trick" carburetor.

The primary reason for choke shroud removal is to provide additional carburetor-to-hood clear-

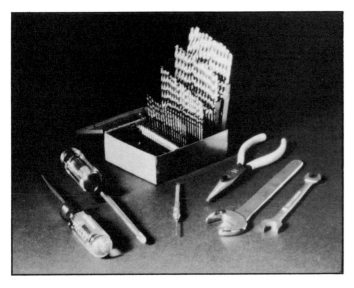

In order to perform the type of modifications described in the text, the proper tools are required. A precision drill set is essential to measure and resize orifices. Suitable tools are generally available under the Craftsman label from Sears.

Milling away the choke shroud (housing) is a popular performance modification. This will increase flow capacity but is most helpful when hood or air box clearance above the carb is minimal. This is best done on a precision mill by an experienced shop.

Filling of flat surfaces on the main body can improve gasket sealing if the body has warped more than 0.010-inch. One of the supposed "carburetor magicians" also chamfered all the holes in the flat surface. This serves no purpose except to increase the chance for fuel leakage.

MACHINING OF FLAT SURFACES
(All Applications)

New carburetors don't require this operation, although most modifiers do it as a means of justifying the cost of their services. The only benefit to be derived from filing or machining the main body flat surfaces (the surfaces which mate to the metering blocks) is improved gasket sealing. It is only suggested if the surfaces have warped more than .010-inch. This may be easily checked with a straightedge. Excessive warpage usually can be corrected using a medium-fine flat file. Machining will be required only for extreme cases.

Another frequently seen cosmetic modification involves chamfering of the holes in the main body flat surfaces. What this is supposed to accomplish is anybody's guess. In fact, it does nothing but impair gasket seal.

Some carburetor shops have also decided to file or mill the flat surface on the metering blocks. Under no circumstance should this be done. The metering block surface contains a number of ridges and two gasket locating pins. The smaller ridges, known as sealing beads, serve to improve gasket sealing. A thick ridge run-

ance so incoming air does not have to turn quite so sharply in order to enter the primary venturis. Minimal airflow increases may be derived from this alteration. While cutting the shroud away (a mill should be used for this operation), care should be taken to make the cut even with the flat surface of the air cleaner mounting flange. It is of the utmost importance that the circular boss around the float bowl vent tube be left intact. After the mill work is complete, all sharp edges should be ground to a 3/8-inch radius wherever possible. A file, high-speed grinder, or electric drill (with rotary file insert) may be used to gently grind this radius. The carb body must be thoroughly cleaned after this operation.

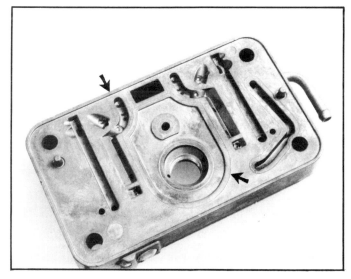

The face of the metering block should never be machined flat. The blocks do not "warp," and milling the surface will remove the all-important ridges that provide a tight gasket seal with the main body.

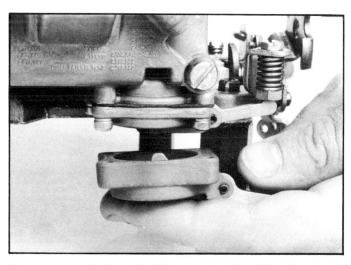

For some applications, the high-capacity 50cc "Reo" accelerator pump kit is recommended. A complete kit consisting of pump diaphragm, housing, linkage, spring, and cams must be used for proper operation. On some types of drag race cars, larger pumps are needed on both the primary and secondary sides (double-pumpers) for optimum performance.

ning around the outer edge of the metering block limits gasket compression. If the block is filed absolutely flat, gasket sealing is considerably impaired. If the float-bowl screws are over-tightened, the gasket may be compressed to the point where it spreads sufficiently to obstruct holes and passages that control fuel flow. This is obviously undesirable as it will alter fuel/air calibrations.

ACCELERATOR PUMP CIRCUIT MODIFICATIONS
(All Applications)

Before considering modifications to this circuit, keep in mind that the purpose of the accelerator pump is to overcome a momentary lean condition when the carburetor is abruptly brought to part or wide-open throttle. The need for an accelerator pump is especially critical when: 1) a drag race vehicle is brought to the starting line with the engine at relatively low RPM (especially critical with an automatic transmission); 2) an oval-track car is brought off a corner; 3) a road-racer is brought out of a low-speed corner.

In order to understand what occurs when the accelerator pedal is suddenly pushed to the floor, picture the fuel flowing through the carburetor with the engine at 3000 RPM and a manifold vacuum of 8.5 inches. Vacuum is drawing fuel through the main metering circuit, and all is right with the world. Suddenly, manifold vacuum drops to zero, so virtually no fuel is drawn into the carburetor bores. Fuel already in suspension falls out when manifold vacuum is suddenly reduced and air velocity through the intake manifold slows dramatically (adding to the lean condition). Simultaneously, the throttle plates are placed in a wide open position which demands maximum fuel delivery. The momentary result is an extremely lean condition (air/fuel ratio of 20-22:1), and engine hesitation or backfire.

This may seem like an inappropriate place for such a lengthy explanation but a good deal of "magic" appears to be associated with the accelerator pump and the effects of modification. In truth, all the pump can do is overcome the momentary lean condition when the carburetor is suddenly cranked open.

In some cases the standard 30cc pump may simply not have sufficient capacity to compensate for the momentary leanness. In classes where changing accelerator pump capacity is allowed, Holley recommends that auto-

Whenever squirter nozzle size is increased beyond 0.042-inch, this hollow screw should be used to ensure adequate flow of fuel to the nozzles. This screw replaces the standard solid screw that attaches the squirter to the main body.

matic transmission cars with double-pumper carburetors use two 50cc pumps.

When changing to the 50cc pump, the pump cam and linkage must also be changed. All necessary conversion parts are contained in Holley kit no. 20-11. In conjunction with the higher capacity pumps, larger discharge nozzles should be used to increase the initial fuel discharge. As noted in Chapter 4, Holley offers three types of discharge nozzles. Nozzle diameter ranges from .025-inch to .052-inch, so the proper size is available for virtual-

Super Tuning and Modifying Holley Carburetors **75**

The most frequent throttle-body swap involves replacing the 1 11/16-inch throttle body of a 660 center-squirter with the 1 3/4-inch throttle body from an 850 cfm double-pumper. This provides an airflow increase of about 40cfm, but the bottom of the venturi bores in the main body must be enlarged to match the larger bores in the 850 throttle body.

When swapping throttle bodies, the diameter and depth of the idle discharge port should also be checked. Many times, the vertical supply passage is not drilled through, so the diameter may not coordinate with other idle system calibrations.

When a throttle body swap is performed, several details must be checked. All idle fuel-delivery passages must be matched and the throttle linkage must be checked to ensure that the linkage mounted on the main body will coordinate with linkage on the throttle shafts. Note here that a slight amount of machine work is required to provide idle fuel delivery from the main body to the passage in the throttle body that feeds the idle discharge port.

ly any application. Determining what is proper is another matter, requiring a good guess to establish a starting point and trial-and-error testing to home in on optimum performance.

If the nozzle is too small, initial acceleration will be marked with hesitation or backfires. The engine may stall from fuel starvation. Conversely, too large a nozzle will result in sluggish acceleration, throttle response will not be crisp, and the car may slow down. A puff of black smoke upon acceleration is another indication that nozzle size is too large. This holds true for either manual or automatic transmission applications.

Standard-shift vehicles generally don't require the large capacity 50cc pumps. In many cases the out-of-the-box accelerator pump system is totally adequate. However, in drag race applications with a relatively low-stall-speed torque converter, a considerable enlargement of the pump nozzles is frequently required. Something in the area of .040-inch to .065-inch should be a good place to start. Any time the nozzle is increased beyond .042-inch, a hollow pump discharge screw (Holley part no. 26-12) should be installed to provide sufficient fuel flow capacity. If a nozzle size of .052-inch or greater is used, the accelerator pump passage in the main body must be enlarged to insure that a full volume of fuel reaches the nozzle. Oval track cars should be able to come off the corners well with comparatively small squirters. In many cases, the nozzle with which the carb was originally equipped is entirely adequate.

THROTTLE BODY SUBSTITUTION
(2x4-Barrel Drag Race)

When using the 0-4224 center-squirter, substitution of the throttle body from an 800 or 850 cfm carb (with 1 3/4-inch bore dimension)

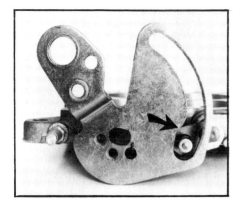

Throttle actuation on a 4224 "center-squirter" isn't 1:1, but it isn't progressive, either. Primaries open about 10 degrees before the secondaries start to open.

One of the advantages of the 8156 (over the 4224) is that the secondary linkage is progressive, allowing 30 to 40 degrees of primary throttle opening before secondaries are activated.

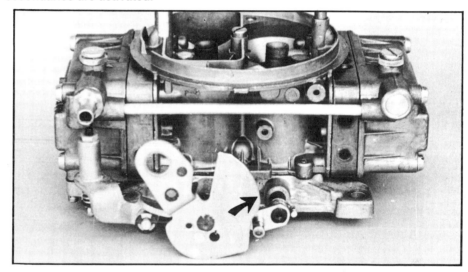

The secondary activation cam of this highly modified center squirter has been reprofiled to delay the opening of the secondary throttle blades. Unfortunately, this modification does not provide for positive closure of the secondaries. This potentially dangerous modification is an example of ill-conceived alterations performed by some carb "specialists."

can increase carburetor airflow by approximately 40cfm. More significantly, enlarging the throttle bore diameter while leaving the venturi diameter unchanged creates a stronger metering signal which starts fuel flowing through the booster venturis earlier in the throttle-opening cycle. When using the 800 or 850 base plate, the bores in the main body must be enlarged and blended to match the 1 3/4-inch diameter bores in the throttle body. This modification, although popular for the 660cfm center-squirter, may not provide the performance levels that can be derived from part number 0-6109, a 750cfm double-pumper designed specifically for dual four-barrel race applications. Some racers prefer modified center-squirters; in installations with carbs mounted inline, it may be the only choice. Part number 0-4224 is of the Model 4160 persuasion, which means it has a secondary metering plate rather than a block. Consequently, it's shorter than a double-pumper and a pair can be installed on intake manifolds with carburetor spacing too close to accommodate a Model 4150.

Irrespective of the carburetor part number, when attempting to marry the throttle body of one carburetor to the main body of another, compatibility is an issue. Many double-pumpers use secondary linkage that pivots around a boss on the main body. When converting a 0-4224 for use with a 1 3/4-inch bore base plate, it is necessary to use a base plate which is fitted with self-contained linkage. The base plates from the 0-4223/850cfm center-squirter (which no longer produced) or the 0-4781-2/ 850cfm double-pumper may be used for this adaptation. Part of the problem created by swapping throttle bodies is that primary and secondary opening rates must be compatible with the accelerator pump system. The 4224 has nearly a 1:1 opening rate (primaries and secondaries open at the same rate). Double pumper-series carbs open progressively (the primaries are partially open before the secondaries are actuated), so the throttle body must be converted to a ratio approaching 1:1.

Failure to perform this conversion will result in an overabundance of raw fuel being squirted into the secondary venturis before the throttle plates are opened. This will lead to one of several undesirable conditions. A volume of fuel may be abruptly dumped into the manifold when the secondaries are opened or, if the throttle plates do not seal especially well against the bores, the raw fuel (delivered to the secondaries by the accelerator pump) will leak

Super Tuning and Modifying Holley Carburetors **77**

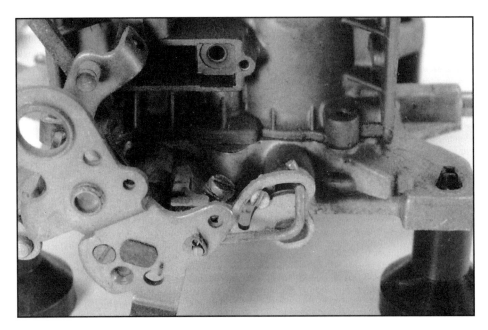

Some old habits die hard. It was once in vogue to install a screw in the secondary throttle lever of vacuum secondary carburetors. When the link from the primary side reaches the screw, the secondaries are pulled fully open. Since there's no accelerator pump on the secondary side, this "poor man's" double pumper routinely results in a serious stumble during the transition to wide-open throttle.

into the plenum before the secondary throttles are opened, causing an over-rich condition. Since this would occur most frequently during the critical starting-line staging time period, consequences could be embarrassing.

An excess of fuel delivered to the combustion chamber at the wrong time could cause a fouled plug or could kill the engine entirely.

Holley kit no. 20-3 may be used to convert any double-pumper to a 1:1 ratio. Holley engineers specifically advise against the use of gear drive units to achieve a 1:1 opening rate. The gear drive places a significant load on the throttle shafts and can cause premature wear and breakage.

NOTE: This is one of the many areas of carburetor modification where one change demands another in order to be effective. As a result, cost is increased significantly, but if the proper carb were originally selected, the same level of performance could be achieved without great expenditures of time and money.

Some carb shops offer modified throttle bodies that have the bores enlarged to the point where the primary and secondary throttle blades contact each other. While this technique can achieve airflows as high as 960cfm, part throttle operation is impaired. The primary throttle plates seal against the secondary plates and manifold vacuum is presented to the secondaries, causing unwanted secondary fuel to flow whenever the primaries are opened. Secondary fuel flow with virtually no airflow through the secondaries creates an over-rich condition at part throttle.

PRIMARY IDLE CIRCUIT MODIFICATIONS
(All Applications)

This is one area where individual tailoring of the circuits can make engine operations significantly smoother. If necessary, the first step is to drill holes approximately 3/32-in. in diameter in all four throttle plates. The purpose of the holes is to increase airflow at idle while allowing the throttle plates to remain in the proper position relative to the transfer slot and idle discharge orifice.

The combination of low-idle vacuum and a relatively high-idle RPM, created by a long duration camshaft, makes for conditions which are beyond the design parameters of the standard circuits. The proper idle RPM may be

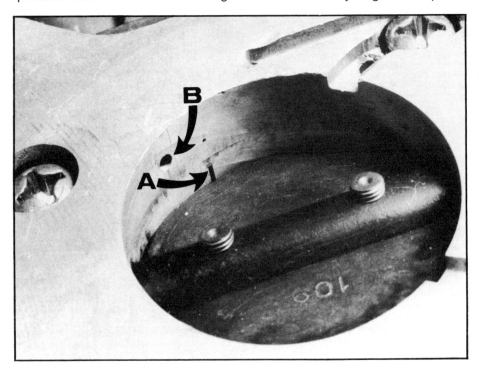

Problems with idle-mixture adjustment quality almost always are traceable to an improper throttle-plate-to-transfer-slot relationship. If, in order to gain a stable idle, the idle-speed screw must be adjusted to open the throttle plate idle position beyond normal limits, the plate may "uncover" too much of the idle transfer slot. This activates flow through the slot (A) and deactivates flow through the idle-discharge hole, which is controlled by the mixture screws.

obtained by adjusting the speed-control screw, but mixture adjustment capability will be lost since the throttle plates may totally uncover the transfer slots. In fact, the plates may be open wide enough to start fuel flowing through the main discharge nozzles. By providing additional airflow capability through the holes drilled in the throttle plates, the desired idle speed may be obtained, yet the throttle plates will cover enough of the transfer slot to provide proper idle mixture adjustability.

Idle airflow can also be increased without drilling holes in the throttle plates. The secondary throttle stop screw, accessible from the underside of the throttle body, can be rotated clockwise one to one-and-a-half turns to increase idle airflow through the secondaries. This method serves to spread the flow of air, yielding an improved balance between primary and secondary throttle bores.

Engines with usually low vacuum may require a carburetor with holes drilled in the throttle plates and increased secondary throttle opening. In all cases the proper idle setting will expose .045- to .060-inch of the idle transfer slot on the primary side, as viewed from beneath the carb. A quick way to check for proper throttle plate position is to monitor the effects of the idle mixture screw. If they have little or no effect on idle speed or quality, the throttle plates are open too wide for proper idle fuel delivery.

In rare instances, it may be necessary to modify the air/fuel ratio of the idle circuit. Methods are described later in this chapter.

SECONDARY IDLE CIRCUIT
(All Applications)

In some instances, a cleaner idle can be achieved if the mixture flowing through the secondary idle circuit can be adjusted. The real need for an adjustable secondary

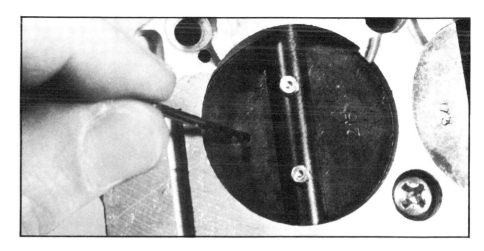

When a high-overlap "racing" cam is used, it also may be necessary to drill holes in the throttle plates to increase airflow at idle. This allows the throttle plates to remain properly positioned with relation to the transfer slot, improving idle characteristics and reducing plug fouling.

For street applications, adjustment of the secondary throttle position (with the stop screw) to increase idle airflow is preferred to drilling holes in the throttle plates. It also should be attempted on race engines when more idle air is needed to alleviate rough-idle problems.

Holley introduced carburetors with adjustable secondary idle circuits in the late 1970s. These types of carburetors (note secondary idle mixture screw), particularly part numbers 0-8156 and 0-8162 (750cfm and 850cfm, respectively), have become standard fare for most hardcore drag and oval-track applications.

Super Tuning and Modifying Holley Carburetors

When an adjustable secondary idle circuit is installed in a double-pumper, this hole must be drilled so that it intersects the vertical hole drilled down from the main body mating surface. Curb-idle fuel will not be discharged through this hole, just as it is on the primary side.

The small hole below the transfer slot is the constant-feed idle discharge, which is used in the secondary barrels of most four-barrel carbs. This hole must be plugged when an adjustable secondary idle circuit is added.

On 0-4781 double-pumpers, this pocket in the throttle body must be drilled to connect the new curb-idle discharge hole in the throttle bore (see above). A throttle-to-main-body gasket can be used as a template.

idle system is sometimes questionable, but it can be advantageous with SOME engines.

Installation of a secondary idle circuit involves several machining operations, some of which differ with the part number of the carburetor being modified. Common modifications include plugging the constant-feed discharge holes in the throttle body and drilling a .090-inch replacement fuel outlet hole in each secondary throttle bore. These holes should be in the same relative location as the primary curb idle discharge outlets. The secondary idle air bleeds will also have to be enlarged to the same diameter as the primary bleeds.

In order to have the secondary idle circuit function as desired, it is necessary to separate it from its counterpart on the primary side. A fuel transfer track machined in the top surface of the throttle body connects both circuits. It must be plugged. This may be accomplished with epoxy, or a small hole may be drilled and an appropriately sized steel plug pressed into place. Some carbs (late-model 850s) may have this track already plugged.

On the 0-4779 (750cfm) double-pumper it will be necessary to open the cored holes in the main body which supply fuel to the discharge holes in the throttle body.

Another required operation when installing an adjustable secondary idle circuit is to block the interconnecting fuel transfer track in the throttle body.

In specialized drag racing applications (especially 2x4 applications), it may be desirable to lean the idle delivery circuit to prevent plug fouling during staging. Here a fine wire has been inserted into the idle feed restriction to reduce the overall area of the restriction.

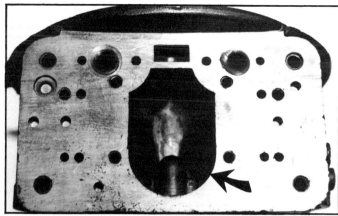

Some racers try to increase float-bowl capacity by drilling through the power valve boss (on metering blocks not fitted with power valves) to let fuel flow into the chamber behind the metering block. It is not necessary to increase the bowl volume, and this procedure is never recommended.

The cored holes in the main body run vertically, rising up from the throttle body mating surface. Two additional cored holes extend from the metering block surface in toward the center of the carburetor. These vertical holes aren't quite deep enough to reach the horizontal ones, necessitating the drilling operation. (Since the entire secondary idle circuit modification amounts to duplicating the primary system, the latter may be used for reference.)

Once each vertical hole is drilled through, an L-shaped passage will be created. Each hole should be blown out with compressed air and inspected for metal cuttings. Then, installation of metering block number 134-153 will make the secondary idle circuit fully operational. However, since the replacement metering block contains a power valve and the original did not, the power valve must either be removed and the hole plugged or the jetting will have to be leaned approximately 6-10 steps to compensate for the additional fuel delivered to the power enrichment circuit. All of this machine work is, of course unnecessary if a carburetor with adjustable secondary idle circuits already included (such as part no. 0-8156) is purchased.

A model 0-4781 (850cfm) double-pumper may be similarly modified for more precise idle mixture control. The approach is identical except that in addition to opening the holes in the main body, two pockets (one for each bore) must be drilled in the throttle body. A throttle body gasket may be used as a template to locate the pockets with respect to each throttle bore. Discharge holes must also be drilled in the throttle body to provide a path for fuel to flow from the pocket to the throttle bore. As with the 4779, the holes in the main body must be drilled deep enough to connect with the cored passages which lead to the metering block surface. Metering block No. 134-212 (with No. 80 or No. 82 jets as a starting point) should be installed to provide secondary idle mixture adjustability. An 850cfm carburetor with an adjustable secondary idle circuit is available under part no. 0-8162.

MISCELLANEOUS IDLE CIRCUIT MODIFICATIONS
(All Applications)

Other modifications to the idle circuits are generally not required, although some carbs, most notably the 660 center-squirters, require leaning of the off-idle air/fuel mixture. The idle feed

The amount of fuel delivered through the power enrichment (valve) circuit is determined by the area of the PVCR in the metering block. If extra enrichment is needed, above what can be obtained from enlarging the main jets, it is possible to enlarge the PVCR area. This is, however, a good way to ruin a perfect metering block. Don't try this unless you are absolutely certain there is no other solution to the problem.

The "picture window" power valve at right has greater flow capacity than either the four- or six-hole version. The high-flow valve without piloting shoulder is the highest flow unit offered by Holley.

Obsolete "nose bowl" is still popular, with oval-track racers who believe that this design offers superior fuel control capability. Evidently Holley didn't agree, since the bowl was dropped from production years ago.

If you are tempted to "modify" the inlet needle-and-seats or the float bowls of a Holley carb, consider this: The twin-turbocharged 454-inch Chevy shown here produced more than 1000 horsepower and over 1000 lb-ft of torque. Fuel delivery was sufficient through a single Model 4500 with unmodified float bowls.

restrictions (normally .028-in. to .035-in. in diameter) located in the metering block on either side of, and in a horizontal line with, the power valve, control off-idle performance. Each restriction is drilled perpendicular to the flat surface and is directly adjacent to a channel hole, which is drilled at a 45-degree angle. (Some metering blocks have the idle feed restrictions positioned in the idle well. With this configuration, alteration of idle fuel mixture cannot be accomplished without extensive drilling.)

The off-idle mixture may be leaned by inserting a 1/2-in. long V-shaped wire, .010-in. to .015-in. in diameter, into the restriction and the adjacent hole. The diameter of the wire will determine the percentage of area reduction and therefore the amount that the mixture is leaned. A fairly stiff type of wire, preferably steel, should be used since the V-shape will be responsible for keeping the wire in place. An alternative is to locate appropriately sized brass restrictions that can be installed and

Super Tuning and Modifying Holley Carburetors

drilled to the desired diameter once the original restrictions are removed.

POWER ENRICHMENT CIRCUIT
(All Applications)

Over the years, Holley has offered several types of power valves, including four-hole and six-hole, and "picture-window" designs. The latter valve is preferable and is available under part no. 125-XX with various ratings between 2.5 and 10.5 inches of mercury (when ordering or buying a power valve, the last two or three digits will designate the opening point — see Chapter 2, Understanding Holley Part Numbers). This valve must be used with a round gasket that has no protrusions on the inside diameter. Ideally, the power valve opening point should be at least 1 to 1 1/2 inches below manifold vacuum at idle. Also available under part number 125-XXX is a higher capacity valve of the "picture window" design. This valve is described in detail in Chapter 4.

Removal of the power valve from the secondary metering block, with a compensatory increase in the secondary main jet size, is an acceptable modification. Engine performance will not be improved, but inspection and maintenance of the secondary valve will be eliminated. However, the primary power valve is essential to clean part-throttle operation while maintaining ample enrichment at full power. If the primary power valve is removed, the hole plugged, and jet size increased the proper amount, air/fuel mixtures at part throttle will be excessively rich, greatly increas-ing the chance of fouling the spark plugs.

When selecting a power valve, the rating should be low enough to insure that the valve is closed at idle, yet high enough to prevent the valve from closing when engine vacuum begins increasing at high RPM. Many high speed leanouts can be traced to a power

Some air bleed modifications use standard Holley main jets as replaceable substitutes for the stock pressed-in brass orifices. In general, it is better to alter the fuel metering system rather than modify the air bleeds.

valve that was incorrectly selected for a particular application.

The amount of fuel flow through the power enrichment circuit is controlled by the power valve channel restrictions, which are visible when the valve is removed. These restrictions vary in size from carburetor to carburetor, ranging from .030- to .120-in. Increasing the size of these restrictions will enrich the air/fuel mixture at full power, but will have no effect at idle or during part-throttle operation (when the power valve is closed). Any increase in power valve channel restriction size should be accompanied by a decrease in jet size if the original wide-open throttle air/fuel ratio is to be maintained.

Altering the diameter of the power valve channel restrictions should be done very carefully. If the holes are drilled too large in diameter it may be necessary to replace the metering block in order to obtain acceptable performance. Fuel flow through the main jet and power valve must be properly balanced for optimum performance under all operating conditions. Keeping the restrictions within .010- to .020-in. of original size will insure that the power enrichment circuit will properly augment the main circuit. In no case should the restrictions be enlarged beyond .120-in., as the power valve flow capacity will be exceeded.

One liability inherent in a power valve that employs a rubber diaphragm is the potential for damage from a backfire. For no apparent reason, some people experience continual problems with power valve blowouts from backfires, while others never do so. With the installation of a power valve check ball kit (part number 125-500) all Holley owners can eliminate the need for periodic power valve replacement.

A popular modification practice by some "professional" carb modifiers is to reduce the section width of the throttle shaft and install thin throttle blades. This modification will only provide minimal airflow increases and it will significantly reduce the strength of the throttle shaft.

The kit includes a drill bit, spring, check ball, and brass seat. Should a backfire occur (after the kit is installed, of course), the check ball is forced against its seat, thereby preventing excessive pressure from reaching and rupturing the power valve.

FLOAT BOWL
(All Applications)

Float bowl capacity is usually of some concern to racers, since most engines have a healthy appetite for high octane gasoline. One method of increasing capacity is to drill the power valve boss in metering blocks not originally designed for use with a power valve. Fuel is thereby allowed to flow into the rather large chamber, between throttle bores, in the main body. This practice is specifically not recommended because the increased fuel volume can overpower the float, adversely affecting operation. The fuel flow capacity of a single center inlet float bowl is sufficient for a 330-cubic-inch engine spinning at 10,000 RPM. Considering that a four-barrel Holley has two such float bowls there should be no need to be concerned about increasing bowl capacity.

Holley offers both steel and Viton rubber fuel bowl inlet needles-and-seats. Viton units should be installed unless the fuel being used causes them to swell, in which case a steel needle-and-seat assembly should be substituted. Steel doesn't have the cushioning effect of Viton, so small pieces of dirt and scratches can cause leakage around the needle, resulting in a loss of fuel level control.

Traditionally, most Holley carbs were fitted with hollow brass floats, but during the past few years, Duracon, a plastic, has largely replaced brass. The two are interchangeable, but as might be expected, plastic floats are less expensive than their brass counterparts.

Solid plastic (Nitrophyl) floats, which are somewhat heavier, are also available, but they provide no performance gains. Being slightly heavier, they possibly respond more slowly to changes in fuel level. Part no. 116-1 fits side inlet bowls and 116-3 fits center inlet bowls. When a center inlet bowl is used in conjunction with the Model 4160 metering block (the one without removable jets) the Nitrophyl float must be used since the brass float will not clear the metering plate. The primary advantage of a Nitrophyl float is that it won't collapse when subjected to high pressures, such as in a supercharged or turbocharged application.

MODIFIED AIR BLEEDS
(All Applications)

In general, both idle and high-speed air bleeds should be left unaltered. The bleed is to air as the jet is to fuel. As bleed size is increased, the air/fuel ratio becomes leaner but size also affects circuit timing. If the high speed air bleed is too large there will be a delay in fuel flow activation through the main metering

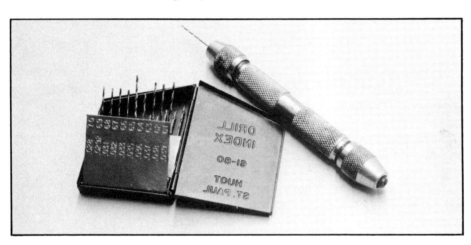

It is possible to alter internal passages of a Holley carb to increase fuel flow for alcohol applications. These alterations require correct tools and careful technique. But it is virtually impossible to enlarge existing passages enough to deliver the volume of fuel required for more than approximately 400 horsepower from a single four-barrel engine. Rather than modifying a gas carburetor, it would be better to switch to one designed specifically to handle alcohol.

Extremely large quantities of alcohol must be metered through the carb to gain suitable air/fuel ratios. Holley offers a selection of needle-and-seat assemblies, jets, and power valves that are suitable for use with alcohol carburetors.

If the main fuel well is to be enlarged in an alcohol application, these plugs in the top of the metering block first must be drilled and removed. The well then can be drilled to a diameter of 0.250-inch, but this must be done carefully to prevent damage to the well wall.

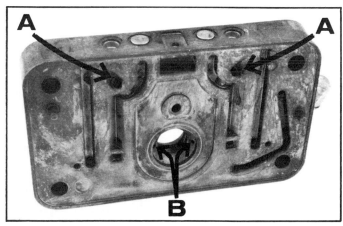

For alcohol applications, the angle passage (A) leading from the main well in the metering block to the booster supply passage in the main body can be drilled to a maximum diameter of 0.180-inch. The PVCR passages (B) also can be enlarged to 0.130-inch maximum diameter.

circuit, which could cause a stumble at part throttle.

It doesn't make much sense to increase air bleed size and then rejet to compensate for it. Some modification companies offer screw-in air bleeds and their easy accessibility may make carb tuning a bit easier. Unfortunately, if you're going to change air bleeds, you must carry a full complement of them, in addition to a complete selection of jets. It is also necessary to reorient your thinking since in air-bleed terminology bigger equals leaner and vice versa.

Alteration of air-bleed diameter can be useful in tuning some engines which may have bizarre manifold vacuum characteristics. In instances where a high idle speed is required, the main metering circuit may be activated at too low an engine speed, thereby causing an overly rich condition. Installation of larger air bleeds can alleviate such conditions. However, if the air bleeds aren't removable, handiwork with a drill bit may cause more problems than it cures. If a larger diameter bleed doesn't cure a particular problem, there's no way to return to a bleed of the original size — unless the bleed is completely drilled out and the boss is tapped to accept a removable bleed.

THINNING OF THROTTLE PLATES AND SHAFTS
(All Applications)

Thinning or "slabbing" of throttle shafts and plates can provide about a 10cfm increase in airflow. All thinning operations must be done very carefully. As material is removed from throttle shafts they become increasingly susceptible to breakage. Throttle plates, on the other hand, won't break under normal operation, but an engine backfire can bend them if they are too thin. It should be noted that the outer circumference of the plate is machined to a precise angle. Any modification which alters the edge of the throttle plate can adversely affect seating in the throttle bore, resulting in poor idle characteristics.

INTERMEDIATE CIRCUITS

Holley employs an intermediate circuit in certain versions of the Model 4500. It is intended to overcome a dip in the fuel-delivery curve of these particular carburetors. Due to the extremely large diameter of the venturis, fuel doesn't begin feeding through the booster nozzles quickly enough, creating a flat spot at part throttle. The intermediate circuit was added to smooth the transition from the idle to the main circuit.

Because few 1050cfm carbs (depending upon the application) need an Intermediate circuit, it stands to reason that smaller 750 and 850 cfm units will perform satisfactorily with only idle and main metering systems. In many cases the addition of an intermediate fuel circuit creates an over-rich condition. This circuit is simply not needed in any Model 4150 or 4160. Save your money.

MODIFICATIONS FOR USE WITH ALCOHOL

Use of alcohol (methanol) in a fuel system originally designed for gasoline creates a host of special problems. But with gasoline supplies constantly being manipulated for political ends, the use of alcohol in competition engines often makes sense. Alcohol also offers some advantages in terms of power output and operating temperatures. Holley offers several carburetors (part numbers 0-9647, 0-9645, 0-9646, 0-80535, 0-80498) calibrated specifically for use with alcohol.

It is also possible to modify existing Holley double-pumper gasoline carbs to handle alcohol. Since creating sufficient overall fuel-handling capability is the challenge in building an alcohol carburetor, calibration is easier with lower air flow capacities. It's

Super Tuning and Modifying Holley Carburetors **85**

also helpful to use carburetors with straight rather than down-leg, boosters (or replace the boosters); find metering blocks without emulsion tubes in the main wells; and be prepared to learn some new "tuning" techniques.

Using alcohol as an internal combustion engine fuel requires air-to-fuel ratios of about 6.5-7.5:1. This is roughly twice as rich as a comparable gasoline-fueled engine (to be precise, it requires 2.4 times the amount of alcohol for proper combustion). Therefore, to make the same amount of power at wide-open throttle, the carburetor must deliver 2.4 times as much fuel! It appears that the simple solution is to drill all of the delivery passages so that the cross-sections are 2.5 times the stock sizes. Unfortunately, there just isn't enough metal around most of the passages to allow such enlargement.

As a practical matter, most stock carburetors don't require 100% fuel flow capacity, even at w.o.t. It is, therefore, possible to gain fairly satisfactory alcohol delivery from a single modified four-barrel to produce about 400-500 horsepower. But (and this is a big BUT), be aware that high speed leanout can be a problem, so go cautiously until you are certain you have enough delivery to prevent damage to the engine.

All Holley metering blocks and main bodies have essentially the same fuel flow capacity. Since the normal fuel volume requirements for alcohol stretches the fuel delivery capacity to the physical limits of the carburetor, restricting airflow minimizes or eliminates the possibility that internal fuel circuit capacity will cause an excessively lean condition at high engine speeds. In other words, when using alcohol in a gasoline carb, the lower the airflow rating of that carburetor, the greater the possibility of achieving a satisfactory fuel-delivery curve.

Use of straight rather than down-leg boosters is necessary, since the fuel passage within the booster must be drilled to approximately .180-in. in order to carry the high volume of fuel required. In fact, a larger diameter passageway would be preferable, but proceeding much beyond .180-in. may weaken the booster casting excessively. The use of straight boosters also insures a more desirable fuel metering signal.

In addition, the fuel flow capability of the metering block must be increased by drilling the main well to .250-in. In order to gain access to the main well, the plug at the top must be removed. A new plug must be pressed into place upon completion of the drilling operation. Various methods can be used to remove the existing plug but the easiest seems to be a multi-step operation that requires a few drill bits, a screw and a pair of vice-grip pliers. (Holley uses special large-capacity metering blocks in its alcohol carburetors.)

The very top of the main well (that area into which the plug is pressed) measures approximately .250-in. in diameter. However, there are several steps in the well wall, moving from top to bottom. At the bottom, by the jet seat, the main well diameter is noticeably less than .250-in. Hence the need for modification. While it might seem

Considering the volume of fuel that must be moved through an alcohol carb, it may be necessary to increase the size of the float bowl. This custom-made large-volume bowl (left) is fitted with dual accelerator pumps.

Drag race engines like this 330-inch Chevy small block produce incredible power with basically stock Holley carburetors. All it takes is knowledge and a logical approach to simple tuning techniques. Holley field representatives attend most national racing events and are readily available to answer questions.

that the most expedient means of removing the plug and drilling the well would be to simply run a 1/4-in. drill down the passage, such is not the case. To assure proper fit when the new plug is pressed in place, the original plug seat should not be disturbed.

Plug removal should begin with the drilling of a hole, approximately 1/8-in. in diameter, in the center of the plug. Then a suitably sized drill may be used to reduce the wall thickness of the plug. Once this is accomplished, the grip of the plug on the metering block will be significantly weakened. A screw may then be threaded into the 1/8-in. hole, and a pair of pliers can be used to pull the screw-and-plug from the well opening. If removal is difficult, an insufficient amount of plug wall material has been removed.

On some metering blocks, the plug is installed down inside the well itself, rather than on top. These blocks may be modified in much the same manner. After the well has been enlarged, a brass cup plug may be pressed into the plug seat. With the well plug safely out of the way, the main well may be easily enlarged to .250-in. As with all drilling operations, this one should be done in a drill press to insure accuracy. In the case of metering blocks equipped with emulsion tubes, it will be necessary to destroy the tube in order to remove it. Unfortunately, Holley doesn't sell replacement emulsion tubes. To replace these tubes after the main well is enlarged, you will have to salvage them from another metering block.

Additional metering-block modifications include enlarging the angle passage that leads from the main well to the booster supply passage (in the main body). This passage may be drilled to a maximum .180-in. diameter. The power-valve channel restrictions should also be enlarged to .125- to .130-in. A high-flow series power valve should be installed in the primary side and No. 95 jets will get the metering circuit in the ball park. It will also be necessary to double the area of the idle-feed restriction.

Several alternatives are available for the secondary circuits. The entire complement of primary modifications may be duplicated and an adjustable secondary idle circuit may be installed as a means of increasing idle fuel supply. With this done, all four bores will function identically. If this does not provide a sufficient amount of fuel, the secondary power valve may be removed completely and the metering jets can be completely removed from the secondary main metering system. With this done it will be necessary to plug the pickup hole that supplies manifold vacuum to the power valve chamber because the chamber will become part of the float bowl. A piece of lead shot or epoxy may be used as a plug.

With the metering blocks appropriately modified, a .045-in. diameter pump shooter and 26-12 nozzle screw installed, and with 6-519 steel needle-and-seats (.150-in. diameter), the carburetor should be ready for a trial run. Lengthy runs at full power should be avoided until it has been verified that fuel supply is ample. A few other considerations involving use of alcohol include:

1) Alcohol is notoriously corrosive! It will attack rubber components, causing rapid deterioration. Gaskets, power valves, accelerator-pump diaphragms and rubber fuel pump and fuel line components must be inspected and replaced regularly. Note — if electric fuel pumps are used, the seals must be checked often. If alcohol gets by the seal and into the armature area, the pump will explode!

2) When using an accelerator pump nozzle .040-in. in diameter or larger on some carburetors, it may be necessary to enlarge the restriction in the passage that supplies fuel to the pump nozzle. This restriction is located beneath the pump-discharge needle valve. The diameter should be large enough to accommodate whatever shooter is being used.

3) Due to the rate at which an engine consumes alcohol, a standard fuel pump will be inadequate. Use of two alcohol-compatible electric fuel pumps, such as Holley VoluMAX part numbers 12-705 or 12-706, is advisable.

4) When rules allow, two small carburetors should be used rather than one large one.

5) Install a 50cc accelerator pump kit (part no. 20-11) on both primary and secondary sides.

Modification of Model 4150/60 carbs for racing is obviously more than a matter of drilling a few holes and some simple machine work. But there is no magic. Straightforward science and logic are the keys to optimum performance. All the previously mentioned mods can be performed by virtually any competent machinist. If you are uncertain about which modifications are required for your specific application, check with a knowledgeable racing carburetor technician, or call Holley's technical department (502/781-9741).

Super Tuning and Modifying HOLLEY CARBURETORS
The Model 4500

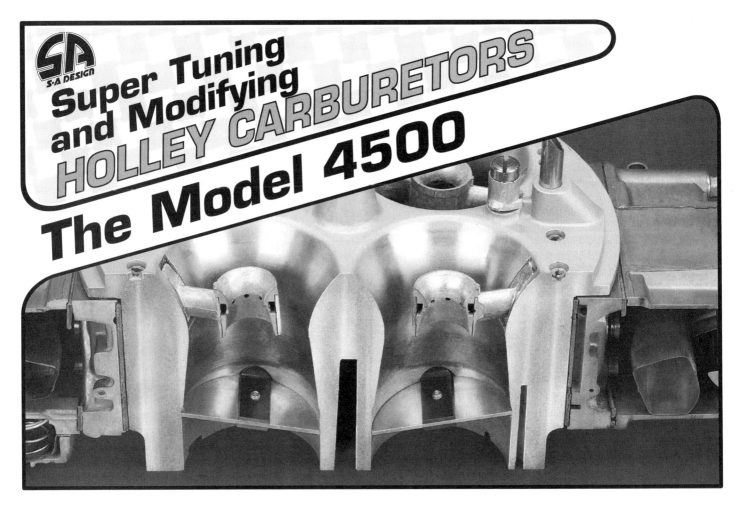

This cutaway view of a Holley HP Dominator shows many of the features that have made these carburetors so popular. This version has replaceable air bleeds, a smooth and contoured main body, high-flow booster venturis, and a reinforced air cleaner stud.

The great factory horsepower wars of the 1950s were brought to a rather abrupt end in 1957 when the automakers agreed to discontinue direct involvement in any type of motor racing. In fact, the agreement did little more than impede developmental efforts by forcing racers to arrive and depart through the rear rather than the front door.

Whether intentional or otherwise, Ford Motor Company has always been adept at producing engines that are entirely capable of living within the spirit of the 1957 accord. In the mid-'60s, when the engines of other auto producers began making sizable gains in race track performance (particularly on the NASCAR Grand National circuit), Ford took steps to equalize the situation. One was to contract with Holley to develop a new four-barrel with superior airflow capacity. To this end the Model 4500 "Dominator" was introduced in 1969.

At first, Dominators were available in limited numbers and only through Ford. Initial use was restricted to Grand National racing and Trans-Am road race competition, and only the top teams had 4500s bestowed upon them. The limited distribution and secrecy that initially surrounded the 4500 led to it being labeled the "Mystery Carburetor." Unfortunately, the aura of mystery has remained to some degree, scaring off many competent engine builders, mechanics, and boating enthusiasts.

A 4500 is actually no more intricate than a Model 4150, so anyone who has experimented with a double-pumper should find a Dominator no more challenging. A considerable number of components used in a 4500 are identical to those included in many 4150 models, and the 4150 and 4500 families are similar in many facets of operation. Stated another way, a 4500 is little more than a 4150 with two-inch throttle bores and a few pieces of modified hardware. The unique features of the 4500 design are the single casting that serves as both throttle and main bodies, the enclosed secondary linkage, and the unique 1/2-in. throttle shafts that are sealed within the carburetor body. Three part numbers within the 4500 family also have intermediate fuel circuits.

Part of the mystery about these carburetors stems from the variety of list numbers (part numbers) offered in the 4500 series. For a number of years there were six individual Dominators. All have 2-in. throttle bores, dual accelerator pumps, center inlet (dual feed) fuel bowls and a unique square mounting flange with studs equally spaced at 5.38 in. Dividing the original series into subsets, four list numbers had venturi diameters of 1 11/16-in. and are rated at 1050cfm; two had 1 13/16-in. venturis and an 1150cfm flow rating. In the mid-1990s, Holley updated the Dominator line, dropped a few part numbers, and added some new ones. The end result is that there are currently eight automotive and three marine Dominators.

The eight automotive part numbers boil down to four different air flow ratings: 750, 1050, 1150, and 1250cfm. Carburetors of the same air flow rating differ with respect to booster style, linkage, and presence or lack thereof of an intermediate circuit. Within the 1050cfm family, part number 0-8082 or 0-8896 is designated for either single four-barrel, or mini-plenum dual four-barrel installations. Whereas the 0-8082 has standard progressive linkage and no intermediate circuit, the 0-8896 has "soft" progressive linkage and an intermediate circuit (making it more suitable for cars with an automatic transmission). Part number 0-9375 is virtually identical to the 0-8896 except that it has annular discharge booster venturis rather than the standard style.

There are also two Dominators rated at 1150cfm (0-7320 and 0-9377) and two rated at 1250cfm (0-80532 and 0-80533). Part number 0-7320 is possessed of standard boosters and progressive linkage, part number 0-9377 has "soft" progressive linkage and annular discharge boosters. Stepping up to the 1250cfm models, the only significant difference between the two part numbers is booster style, which is standard for the 0-80532 and annular discharge for the 0-80533.

Marine Dominators are available in 1050, 1150, and 1250cfm ratings with little else to differentiate one carburetor from the others.

With the advent of the Holley Pro Dominator ram manifold, it became practical to run a pair of Dominator carbs on a small-block Chevrolet. Look closely and you can see that this racer is a bit confused and thinks carburetors are designed to hold spark plugs during impromptu tuneups.

The two distinct types of Dominators — with and without intermediate fuel circuits — are shown here. All three bleeds of each venturi on the carb at left are fitted with brass restrictions, indicating an intermediate circuit is used. On the right, only two bleeds (per venturi) are fitted with brass restrictions, indicating the intermediate circuit is not functional. However, there is one exception. The 8896 has only two bleeds fitted with restrictions and one blank bleed (per venturi), but it does have an intermediate circuit. In this one exception, bleed air is admitted through the unrestricted bleed, and fuel control is established by a restriction in the intermediate fuel-delivery tubes in the venturis.

Part number 0-80340, rated at 1050cfm, has standard boosters and no intermediate idle system; part number 0-75010, rated at 1150 cfm, has annular discharge boosters and an intermediate idle system. Part number 0-75011 is essentially a 1250cfm version of the 1150 cfm model, but has standard progressive rather than soft progressive linkage.

Although the Dominator was designed strictly for racing, it is occasionally installed on a street vehicle or recreational boat. Such use is a mistake 99% of the time; a 454 cubic inch powerplant would have to turn 8000 RPM in order to use the airflow capacity of a single 1050cfm carburetor. Recognition of that fact, combined with the fact that no other carburetor has a Dominator's under-hood presence, led to the development of the 0-80186, the 750cfm version of the Dominator.

When dealing with one or more Dominators, air-flow capacity is always a consideration. Even some full-tilt race engines cannot

One of the advantages of the Dominator design is the fact that primary-to-secondary linkage is internal. Throttle shaft openings are sealed to prevent dirt from interfering with carb operation.

fully use the flow capacity of a pair of 4500s. Others can fully use as much as can be supplied by six Dominators. However, "over-carbureting" is a far more common problem than "under-carbureting" and more often than not, selecting a "big carb" results in a "big problem." As always, the best all-around performance is achieved when air-flow capacity is matched to engine demands. That points to making a determination as to which Dominator best suits a particular application. Starting with the proper carburetor will reduce the amount of modification necessary to accommodate the eccentricities of a particular engine/chassis or engine/hull combination.

As a general rule, part number 0-6214 (1150cfm, no longer available, but previously owned models are still abundant) can be eliminated from the selection process unless the requirement is for a pair of carburetors for a large-displacement engine (427 cubic inches and larger) equipped with certain intake manifolds. The 6214 was designed for Chrysler Hemi individual runner (IR) applications and requires some amount of rework for use on single four-barrel and plenum-type 2 x 4 manifolds. Installation of the standard booster venturis (which are shorter and larger in diameter than the boosters with which the 6214 is originally equipped) is recommended when this carb is used on a big block with a large-plenum, dual four-barrel manifold. (Typically, the intermediate circuit is too rich for non-IR installations and must be restricted as well.)

Other list numbers, such as the 6464 (no longer available new) may be used on most 2x4 plenum manifolds with little or no modification. The 6464 offers 1:1 throttle linkage, an intermediate idle system (as does the 6214) and a 1050 airflow rating. This carburetor was calibrated for use on 350-405 cubic inch engines with 2x4 ram-type manifolds. The 6464 may also be used with great success in single four-barrel applications where an automatic transmission and torque converter are employed. The intermediate system seems to aid starting line launch. (See intermediate system description for pertinent modifications.)

The intake manifold technology of the late 1970s, seen in the Pro Stock/Modified Production/Competition Eliminator classes suggested a decrease in plenum volume (compared to older designs) which in turn demanded specific carburetor calibrations and operational characteristics. Part numbers 0-4575, 0-7320, 0-8082, and 0-8896 were the recommended carburetors for various mini-plenum 2x4 applications, but they have also enjoyed great success in specific single carburetor installations.

For a number of years, the 0-4575 was the only Dominator offered with a choke mechanism. (The 8082 originally contained a choke, but it was eliminated through a production change.) It also has the distinction of being

A close look at the throttle-bore diameter reveals that this is no ordinary Model 4500. Unofficially, it's a 4600 with 2 1/4-in. (rather than 2-in.) throttle bores, developed in conjunction with Holley's individual runner manifold. Problems with reversion could not be eliminated and the project was dropped.

the only 4500 without an adjustable secondary-idle circuit. Unless a choke is required, the 0-8082, which is essentially a 4575 with an adjustable secondary-idle circuit (and no choke, of course), is a better choice when a 1050cfm carb is required. Circle-track racing was about the only area where the 4575 enjoyed any popularity. It is no longer included in the Holley catalog, but the 0-8082 is still available.

When Holley developed the "Pro Dominator" 2x4 ram manifold, the engineering department also engaged in extensive carburetor experimentation. The intent was to tailor a 4500 for optimum performance when installed on the new intake manifold. The result of this effort was the 0-8896, basically a 6464 with special "soft-staged" linkage and a recalibrated intermediate circuit.

Part number 7320 is an 1150cfm carburetor with adjustable secondary-idle circuit, progressive linkage, and metering bodies machined for power valves. This carb is used extensively in single, four-barrel "Econo" drag race classes, and occasionally on 2x4 mini-plenum, ram-type manifolds.

It should be obvious from the above descriptions that each of the part numbers in the 4500 series was developed to meet specific engine requirements. In spite of the magic and mystery purveyed by the self-proclaimed experts in the "carburetor modification" business, extensive modifications are not required if the correct part number is selected at the outset. It doesn't make sense to buy a carburetor, pay a few hundred dollars for "custom modifications," and end up with essentially the same thing that could have been purchased under a different list number. This is not to say that all carburetor modifications are unnecessary. Performance improvements can be realized through tailoring, but

Manifolds with Dominator-type mounting flanges have never been widely available, so many custom designs have been produced. This model, similar to the small-block Chevy "Smokey Ram," was made by Bud Moore and went into Cleveland Fords.

The side view of Model 4500 shows a similarity of primary and secondary fuel systems. Except for the Model 4575, all Dominators have adjustable seondary-idle mixture capacity.

butchering should be left to the people at the meat counter.

With the advent of the redesigned 4500 carburetors, now listed as Dominator HP carbs, the need for modifications is even less. All HP Dominators feature a 1/4-in. taller main body and venturi, machined main body surfaces, contoured venturi entry, reinforced air cleaner stud and vent tube bosses, replaceable air bleeds, high-performance booster venturis with greater fuel flow capacity, revised metering blocks with longer emulsion tubes and larger fuel passages, raised accelerator pump nozzles (to minimize fuel pullover at high RPM), and .120-inch diameter needle and seat assemblies. Except for some minor calibration work, the HP series Dominators are race ready right out of the box. However, if one of these carburetors isn't within your budget, or if you have an older version, here are some of the more productive 4500 modifications.

AIR HORN EXTENSION

Little airflow increase can be attributed to removal of the air horn extension per se. On many race cars, this modification is useful as a method of increasing the carburetor-to-hood clearance. Additional clearance does not increase airflow through the carburetor; it merely insures that a sufficient amount of air will reach the carb inlet. However, elimination of the air horn extension does remove unnecessary material and simplifies some tuning operations.

Super Tuning and Modifying Holley Carburetors

4500 SPECIFICATIONS BY PART NUMBER

LIST NUMBER	CFM	ACCEL PUMP	CHOKE	POWER VALVE	IDLE ADJ.	BOOSTER VENTURI	INTER MEDIATE	THROTTLE DIAMETER	VENTURI DIA.	JETS	PVCR	IDLE FEED	PUMP NOZZLE	LINKAGE
0-4575	1050	BIG	YES	YES	2	SHORT	NO	2.0"	1-11/16	84	.0935	.042 / .041	.035 / .035	STAGED
0-6214	1150	SMALL	NO	NO	4	LONG	YES	2.0"	1-13/16	95	.0935	*	.026 / .026	1 TO 1
0-6464	1050	BIG	NO	NO	4	SHORT	YES	2.0"	1-11/16	88	.0935	*	.035 / .035	1 TO 1
0-7320	1150	BIG	NO	NO***	4	SHORT	NO	2.0"	1-13/16	95	.0935	.042	.031 / .035	STAGED
0-8082	1050	BIG	NO	NO***	4	SHORT	NO	2.0"	1-11/16	84	.0935	.042	.035 / .035	STAGED
0-8896	1050	BIG	NO	NO	4	SHORT	YES***	2.0"	1-11/16	88	.0935	*	.035 / .035	SOFT STAGED

* .025-inch flowed tube in main well
** Restricted
*** Metering block machined for power valve

This chart compares the characteristics of various Model 4500 carbs. As with all carburetor applications, the first and foremost consideration is the selection of the correct carburetor for the specific application under consideration.

When the upper air horn material is machined away, care should be taken to avoid cutting into the air-bleed bosses, fuel bowl vents, or accelerator pump shooters. Machining should be done on a mill so that sensitive areas are not inadvertently cut. Additionally, the entry into each throat should be smoothed and blended with a 3/8-inch radius for minimum disruption of entry airflow.

Holley offers velocity stacks that improve airflow and give the 4500 an even more impressive appearance.

MACHINING FLAT SURFACES

As with the 4150/4160, this modification is usually unnecessary. See Chapter 8 for a complete description.

ACCELERATOR PUMP

With the exception of the 0-6214, all Dominators are fitted with the 50cc large-volume accelerator pump. This pump functions exactly like the smaller version, but the extra fuel volume delivered with each stroke necessitates use of an entirely different pump cam. Several 50cc pump cams are now available, and the yellow and brown cams enjoy the most widespread use.

Since Dominators are equipped with hollow-discharge nozzle screws, there is no need to change the screws with an increase in nozzle size. Replacement screws may be ordered under part number 26-12. Chapter 8 contains a complete description of the accelerator pump circuit as it applies to race applications.

FLOAT BOWL

All Dominators are fitted with "dual feed" center-inlet fuel bowls identical to those used on the 4150 series. See Chapter 8 for a description of pertinent modifications.

MAIN METERING SYSTEM

When a 4500 (or a pair of them) is feeding a race engine operating at high RPM, a healthy volume of fuel must flow through the carburetor. In rare instances fuel flow requirements may exceed the capacity of the No. 100 jet, the largest size offered by Holley. One method of increasing overall fuel delivery capacity is through installation of power valves in carburetors not originally so equipped. The additional fuel

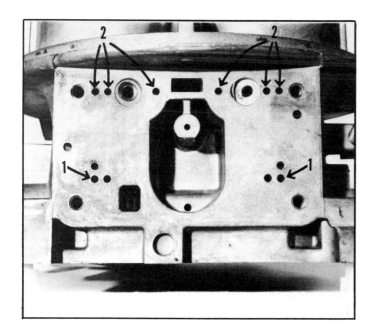

From the metering block side, the main body does not appear much different from a double-pumper. Idle fuel enters the main body at the same place (1). Air bleeds (2) are similarly placed.

The annular discharge booster design is less restrictive than the standard design, providing a stronger signal for the main system. It allows the main system to "start up" earlier and substantially decreases mid-range "lag" caused by the large venturis of the Model 4500. This relatively new design eliminates the need for an intermediate circuit (in most cases) and is a standard feature on some late-model 4500s.

metered through the power enrichment circuit boosts total fuel flow approximately the equivalent amount as increasing jet size eight to ten numbers, depending upon the power-valve channel restriction size. This is the reasoning that led to the use of plugged, rather than unmachined, power-valve mounting holes in the metering blocks of 7320 and 8082 carbs. Should the need arise for increased fuel-flow capacity, the plugs may be removed and replaced with power valves. The power valve channel restrictions are already drilled.

A modified-flow metering block kit (part no. 34-8) is available to boost the fuel handling capabilities of the 6124, 6464 and 9377 carburetors. These carbs are equipped with an inter-mediate system and therefore use a unique metering block. No provision is is made in the original blocks for installation of power valves. Conversely, the metering blocks contained in the modified flow kit are drilled to accept power valves and offer additional fuel-flow capability by virtue of their enlarged main wells.

Modification of the high-speed air bleed diameter appears to be one of the steps generally taken by carburetor modifiers to insure an extremely high "confusion factor." Typically, the air bleeds are enlarged over the standard dimensions. This has two effects: the mixture is leaned and activation of fuel flow through the booster venturis is delayed. In some applications, especially where carburetors with intermediate systems are employed, enlarged high speed bleeds may offer a performance improvement. However, for the most part, it is best to leave the air bleeds unmodified until all problems even remotely connected

Taking a manifold's-eye view, carburetors not fitted with intermediate systems have only a curb idle discharge hole (1) and idle transfer slot (2) visible. Carbs with intermediate metering capability have an additional tube (3) pressed into the throttle bore at the base of the venturi. This is the tube through which intermediate fuel flows.

Super Tuning and Modifying Holley Carburetors

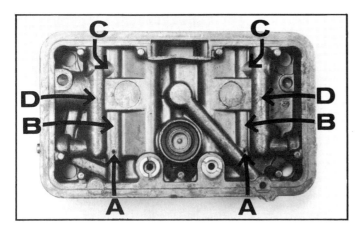
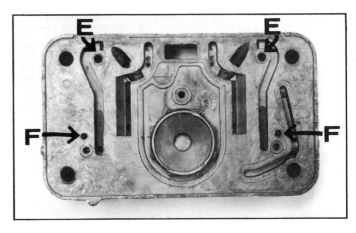

Metering blocks with an intermediate circuit differ from standard. Intermediate fuel enters the circuit through a small hole (A), travels up the well (B), across at (C), and down (D) to exit the hole on the opposite side. Changing the jet size has no effect on intermediate circuit.

Bleed air is introduced at hole (E) and an emulsified mixture exits at (F) into the main body, where it is delivered into the intermediate tubes pressed into the venturis.

INTERMEDIATE SYSTEM

Part number 6214, the 1150cfm carburetor originally developed for IR manifold configurations, was the first 4500 to be fitted with an intermediate system. Inclusion of this circuit was required to affect a smooth transition from the idle to main metering system. With a 1-13/16-in. venturi diameter and a booster venturi of comparatively small cross-sectional area, the 6214 develops a relatively weak fuel delivery signal at low engine speeds. The 1:1 throttle linkage further causes a reduction in the signal because total airflow is divided among eight (on a dual four-barrel installation), rather than four, throttle bores. (Four would be the case with progressive linkage, where there is virtually no airflow through the secondaries until well into the throttle-opening cycle.) This situation results in a fuel delivery "hole" as an engine is brought off idle because airflow demands more fuel than can be metered through the idle system, yet airflow is insufficient to activate fuel flow through the main nozzles.

The intermediate system performs the function that the name implies, filling the gap between the idle and main circuits. Since the system discharges fuel at the base of the venturi, intermediate fuel flow continues at all post-activation throttle openings, including wide-open throttle. Depending on the diameters of the air bleed and discharge tube, the intermediate system can enrich the mixture up to one full air/fuel ratio.

Part numbers 0-6464 and 0-8896 are also fitted with intermediate

with carburetion and intake manifolding have been solved. Then it might make sense to experiment with high-speed bleeds in order to extract absolutely optimum performance. Gains here require a dyno, extensive drag strip testing, and thorough knowledge of 4500 carburetors.

Idle fuel is metered through a restriction in a tube that resides in the main well.

Idle tubes may be seen in main wells when jets are removed.

Two small holes with lead plugs serve to identify intermediate circuit metering blocks. These blocks are sometimes used on highly modified 4150s which have had an intermediate circuit added.

94 *Super Tuning and Modifying Holley Carburetors*

systems, although the latter carburetor has markedly different calibrations. Considering that the 8896 was specifically developed for the Holley 2x4 Pro Dominator intake manifold, it should be usable in any mini-plenum application with only minor adjustments. On the other hand, the 6464 discharges intermediate fuel through a .095- to .100-in. diameter tube and this frequently results in an excessively rich mixture. A .046-in. restriction pressed into the discharge tube will generally cure the overrich condition. Another means of leaning the intermediate system is to remove the drilled brass intermediate air-bleed restriction. Restricting the discharge tube and opening the air bleed will provide a calibration very near that of the 8896.

AIR BLEEDS

There is some confusion about the identification of air bleed restrictions in the 4500 main body. Although the high speed bleeds are always the ones closest to the fuel bowl vent tube, the idle air bleeds are relocated in 4500 carburetors that contain an intermediate circuit.

On carburetors that have no intermediate system, the idle air bleed is located on the inboard side of the D-shaped boss adjacent to each venturi. In these instances the outboard hole does not have a brass restriction in place. Conversely, when an intermediate circuit is present, the outboard hole becomes the idle air bleed and the inboard hole becomes the intermediate bleed. List numbers 0-6214 and 0-6464 have brass restrictions in each bleed boss, but the 0-8896 has no brass restriction in the intermediate hole. As mentioned previously, the brass restriction was eliminated as a means of increasing bleed size, which in turn leans the intermediate mixture.

This close-up of the air bleed area clearly shows the unused bleed holes (1) in a non-intermediate circuit equipped 4500. Bleeds closest to fuel bowl vent (2) meter main system (high-speed) air while outer restrictions (3) handle idle air.

POWER ENRICHMENT

In the 4500, as well as the 4150/4160 series, fuel flow through the power enrichment circuit is controlled by the size of the power-valve channel restrictions (PVCR). In the metering blocks used on 4500 carburetors, the PVCR's measure .0935-in. and represent potentially greater flow capacity than offered by standard power valves. For just such situations, Holley manufactures higher capacity power valves listed under part numbers 125-XXX. Externally, there is no difference between the standard and high-flow power valves. Flow capacity in the high-flow series is greater because there is no internal pilot shoulder to keep the valve stem aligned. The increased clearance between the valve stem and body allows for greater fuel flow. But valves without a piloting shoulder are subject to incomplete closure should a piece of foreign matter find its way to the seat area.

LINKAGE

Operation of the secondary throttle shaft is controlled by a three-piece linkage which is tucked between the throttle bores. Three types are available, 1:1 (direct), progressive (or staged), and soft-staged. The 1:1 version, used only

On intermediate-equipped carbs, the situation changes slightly. The idle air-bleeds are moved to the outermost positions and the intermediate bleeds (2) are installed in the positions between the high-speed (3) and idle bleeds.

on 6214s and 6464s, may prove to be troublesome for some applications, creating low-speed driveability and staging problems. In other instances, it may be partially responsible for a poor launch because the car will have to be brought to the line with all throttle bores open.

The staged linkage delays secondary opening until the primary throttles are open to approximately the half-throttle position. Therefore, since only half the carburetor is in use during staging, air velocity is higher. This can sometimes make for a harder launch because the higher air velocities during staging help smooth the sudden transition to wide-open throttle. On an individual runner

Super Tuning and Modifying Holley Carburetors

The linkage that activates the secondary-throttle shaft is tucked inside the main body between the throttle bores.

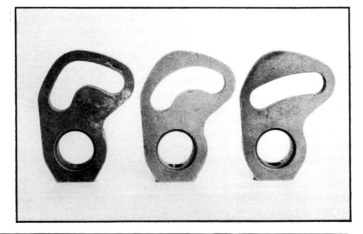

Three types of throttle links are available for the Model 4500. At right is the 1:1 linkage (part number 20-5), center is the "soft-staged" progressive linkage (part number 20-20), and at left is the standard progressive linkage (part number 20-6).

If the idle mixture is consistently too rich on racing engines fitted with dual 4500s, it is possible to lean the idle delivery circuit by reducing the area of the idle feed restrictions. This is most easily accomplished by inserting V-shaped wires into the idle feed restrictions. This is the same technique as described for center-squirters.

manifold, 1:1 linkage is the only choice, but there are other applications where the direct link is preferable.

During the development of the 8896 carburetor, Holley engineers found that there the existing 1:1 nor the staged internal linkage offered optimum performance in 2x4 applications. One of the biggest problems with the staged linkage was the excessive strain placed on carburetor-to-gas-pedal connections when the secondaries were opened quickly. Since the primaries are almost halfway open before the secondaries are moved, the opening rate of the secondaries must be extremely quick in order for both throttle shafts to reach wide open simultaneously. The extreme linkage load created by sudden secondary activation may be high enough to bend or break the pedal linkage in a 2x4 installation.

Wishing to alleviate the driveability and staging problems associated with 1:1 linkage, Holley engineers designed the "soft-staged" link for the 8896. This linkage is now available as a separate item. It allows 15 degrees of primary throttle movement before initiating secondary opening. Since the opening of the secondaries starts sooner than with the standard progressive linkage, the transition to wide open throttle is smoother and creates less strain on the throttle linkage.

BOOSTER VENTURI

Three distinct types of booster venturis have been used in Model 4500 carburetors. The large diameter, short-length booster was used in all Dominators except the 0-6214, which is fitted with a small diameter booster of considerably greater length. The latter booster was developed for individual runner manifolds. The additional length serves only to contain the fuel "stand-off" which is characteristic of IR manifolding. Although the longer booster may look like the Trick-of-the-Week, it should not be used in any single four-barrel or mini- or deep-plenum 2x4 manifolds.

Having a smaller cross-sectional area than the standard, the "long" booster does not offer as strong a fuel delivery signal. Ill-advised racers who have tried the long booster swap have almost universally experienced severe engine stumble during starting-line launch. As mentioned previously, the only worthwhile booster change involves installation of the short booster in 6214s used on non-IR manifolds.

IDLE FEED RESTRICTIONS

Two types of idle-feed restrictions are used in the Dominator series. Models not fitted with an intermediate system meter idle fuel through drilled brass restrictions pressed into the metering body. This is the same arrangement used in most double-pumpers. Carburetors fitted with intermediate circuits admit fuel into the idle system through tubes that reside in the main fuel wells.

If the idle mixture is consistently too rich, a piece of small-diameter wire may be inserted into the drilled brass restriction and bent around into the adjacent hole in the metering block. Typically, a .010- to .015-in. wire will sufficiently lean the idle mixture. Care should be taken that the wire is long enough and bent at an angle that will prevent it from coming loose during operation. An alternative to the wire is to remove the brass restriction and replace it with one that is drilled to a smaller diameter.

At this point, it should be obvious that the Holley 4500 is really not the "mystery" carburetor it's reputed to be. Earlier chapters explained operation of the Model 4150/4160 carburetors in great detail. Much of the information presented in this chapter ties directly to data contained in those sections. Solving the mystery of the 4500 is therefore a matter of reading and understanding the information presented in this book.

The booster venturi on the left was once thought to be a "speed secret" because it was relatively rare. In fact, it was designed for Individual Runner (IR) manifolds, and the only purpose of its extra length was to contain fuel standoff.

Sometimes, even two Dominators aren't enough. This Jon Kaase-built Ford "Boss" engine displaces over 800ci and uses three Dominators on a custom-built sheet-metal manifold. To improve fuel distribution, the Dominators are cut in half and each side is positioned directly over the runners.

Super Tuning and Modifying Holley Carburetors

Super Tuning and Modifying HOLLEY CARBURETORS
Tune-Up Tips for Economy & Special Applications

ECONOMY

During the past few decades, volatile gasoline prices and supply conditions varying from shortages to excesses have become commonplace. As such, concerns over fuel economy range from nonexistent to my-God-what-are-we-going-to-do-the-price-of-gas-just-went-through-the-roof. In fact, fuel economy should always be a concern because excess fuel consumption not only increases pollution levels, it's wasteful and can lead to engine damage. Proper carburetor tuning techniques can improve the mile-per-gallon figure achieved by any vehicle. However, optimum fuel economy can be obtained only through a coordinated effort involving modifications of the carburetor, intake manifold, exhaust system, ignition system, and drivetrain.

As an example, a Third Generation Corvette with which I am acquainted derives power from a 372 cubic inch small block that produced over 420 horsepower during dynamometer testing. The car runs consistent 12.8- to 12.9-second quarter mile elapsed times (at over 110 miles per hour) with standard radial tires and mufflers in place — yet achieves over 21 miles per gallon on the highway.

The premise of this particular combination is that a high horsepower engine can "loaf" under load conditions that would strain an underpowered weakling. Since the engine is rarely stressed, manifold vacuum is consistently high, providing for efficient fuel vaporization. With all due respect to the efforts of the automakers, it should be noted that emissions control equipment on the engine in question is limited to a PCV valve. However, it should also be mentioned that the 372 was not built primarily with gas mileage in mind. Were that the case, figures in the 25 mpg range might be possible

Regardless of make, year, or model, one of the first steps to improved fuel economy should lead to a Sun distributor machine. Most standard centrifugal advance curves are designed to accommodate the lazy driver who lugs the engine or neglects to downshift when necessary. Increasing spark advance at low engine speeds will cause detonation under such operating conditions. However, assuming that other engine parameters allow (cam, compression, etc.), initial advance may be moved to 14-16 degrees before top center and the centrifugal curve may be modified to provide an additional 20-22 degrees at 2500 engine RPM. The optimum advance curve for each engine will be dependent upon such things as combustion chamber shape, connecting rod length and gasoline octane. However, these settings are a good general baseline and further alterations will be minimal.

In addition to centrifugal advance, some sort of vacuum spark advance must be included for optimum efficiency under cruise conditions. With a quick centrifugal curve, most standard vacuum-advance mechanisms will move timing into the pre-ignition area whenever engine speed is in the 2000-2500 RPM range and manifold vacuum is above 10 inches. This may be corrected by placing a restrictor and a bleed in the vacuum line to delay activation. Another option is to install an adjustable vacuum advance canister that allows the vacuum advance curve to be altered. As a final ignition modification, a high output electronic system that eliminates the points should be installed if the engine is not already so equipped.

With increased combustion efficiency, elimination of exhaust gases becomes more critical. Headers have long been touted as mileage improvers. However, the conventional large-diameter, individual-tube configurations do a lot more for performance than for economy. If mileage is of paramount concern, smaller diameter tubing (1-3/8 to 1-1/2 inches) and increased length (45-55 inches) is most desirable. They should be connected to a pair of free-flow mufflers through 2-1/2 in. diameter exhaust pipes.

Assuming that no other major changes will be made, carburetor jetting may now be altered to complement the ignition and exhaust system rework. Theoretically, this means running the air/fuel mixture as lean as practicable. While it may seem incongruous, it is possible to make a mixture so lean that mileage deteriorates. When this point is reached, driveability drops off so much (due to fuel starvation) that the driver is constantly stabbing the accelerator pedal toward the floor, causing a reduction of manifold vacuum and unnecessary power valve opening. Enriching the mixture slightly allows a reduction in throttle opening and maintenance of higher manifold vacuum.

However, a jet size that may be optimum for cruise conditions may also be too restrictive for light loads. When this situation is encountered, a two-stage power valve may be the solution. The first stage, opening as high as 12.5 inches, can eliminate ping or lean-stumble under partial load, while allowing cruise mixture to remain as lean as possible. If leanness is a problem only under full load, the power valve channel restrictions may be enlarged to compensate. However, with a two-stage power valve, the PVCR diameter should be limited to a size that correlates to the maximum flow capacity of the valve (.060-inch).

On some carburetors, reduction of the idle feed restriction diameter is a useful modification. On metering blocks with exposed restrictions a V-shaped wire may be inserted and staked back into the adjacent hole that leads to the main well. Wire diameter must be computed based on the area reduction required. The necessary formulas are contained in Chapter 6.

Alterations made in the idle circuit will be most noticeable when low-speed cruising predominates. Remember, fuel for such operation is delivered through the transfer slot, which is part of the idle circuit. Creating a "leaner" part-throttle fuel delivery to the transfer slot through idle feed restriction or air bleed alterations can be touchy business. This same circuit delivers curb-idle fuel, and reducing fuel delivery to the transfer slot will also reduce delivery to the curb-idle port. It is possible (easy!) to reduce delivery so much that the engine simply will not idle.

Be it jetting, idle circuit modifications, or power enrichment system alterations, tuning a carburetor for maximum economy must be an individual effort designed to optimize fuel metering characteristics for particular operational parameters. There are an infinite number of variables, a fact that precludes establishment of singular "cure-all" procedures. The intake manifold is certainly one of these.

In general, dual-plane designs offer some potential mileage advantages because they are constructed for maximum efficiency at relatively low engine speeds (compared to single-plane designs, which have larger plenums and "straight" runners). A dual-plane manifold produces excellent low-speed torque and frequently requires smaller throttle opening than a single-plane manifold to handle the same amount of load. This results in higher intake manifold vacuum which enhances fuel vaporization. It would be safe to say that a dual-plane manifold would be most effective when engine RPM is predominantly in the 2500-3000 RPM range and manifold vacuum is above 16 in/Hg.

The Model 4360 was Holley's lowest cost four-barrel, and is a direct Quadra-Jet replacement. Although this carburetor was discontinued many years ago, it's still an excellent choice (if you can find one) when fuel economy is of paramount importance.

Before any changes for increased economy are made to the carburetor, the ignition must be in tip-top shape. A Sun distributor machine is absolutely essential to ensure proper distributor operation and to custom-tailor the spark curve.

For maximum economy with a street engine, the ignition must be a fully operational vacuum-advance system. Modern breakerless distributors will operate to 8000 RPM, even with vacuum advance mechanisms, and the extra advance provided by the vacuum during part-throttle operation will increase fuel economy greatly.

In conjunction with an electronic triggering device, a high-output system should be installed. A Holley Annihilator ignition system minimizes the possibility of misfire, thereby improving fuel efficiency. It also will keep the carburetor happy because it's a relative.

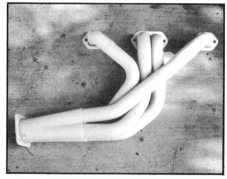

Header manufacturers long have been touting the advantage of their products, but performance models use a tube diameter that is often too large for many late-model engines. "Mileage" headers are generally fabricated of 1 3/8- or 1 1/2-in. tubing.

In some cases, a plenum divider may be used to improve the low-speed performance and economy characteristics of a single-plane manifold. Although a divider will intensify vacuum signal, resulting in a richer air/fuel mixture and a decrease in fuel economy, the carburetor can be rejetted and the idle mixture adjusted, resulting in an improvement in fuel efficiency.

The Holley "Z" Series single-plane manifolds, designed in the 1970s by Zora Arkus-Duntov (father of the Corvette), use a divided plenum in combination with a balance tube for strong low-end performance. The "Z" manifold was engineered to combine the most favorable aspects of existing single- and dual-plane technology. However, this manifold produces an exceptionally strong vacuum signal which points up the need to match a carburetor with the manifold on which it is installed. A carb that is calibrated for a low-signal manifold may run excessively rich when installed on a Holley "Z Series".

Since manifolds and carburetors are intimately involved in the metering and distribution of fuel, they quite obviously exert an influence over fuel economy. The effects of the drivetrain are not so readily apparent. Consequently, an engine that slowly sips its ration of gasoline may seem to be a guzzler simply because certain driveline components are inadequate.

A slipping torque converter and/or automatic transmission are the most common mileage pirates. At best, these devices are 98% efficient in their ability to transmit power. With the passage of time, slippage can only increase, causing a decrease in efficiency. The result is a higher engine speed for a particular road speed.

With the most commonly found rear axle gearing, an engine must turn approximately 1,000 revolutions per minute for each 25 miles per hour of vehicle speed. (This assumes a transmission without overdrive.) This translates to:

SPEED (MPH)	ENGINE (RPM)
50	2000
55	2200
60	2400
65	2600
70	2800
75	3000

When the transmission is operating at 98% efficiency, an additional 45 engine RPM would be required to bring the driveshaft up to 2000 RPM, the speed necessary to maintain 55 miles per hour in high gear. However, when efficiency drops, engine speed must be increased significantly in order to maintain desired speed—and that seriously impacts fuel economy.

These efficiency calculations are theoretical, with no consideration given to tire slippage or other influencing factors, but they do establish a reference base. As an example, with only 80% through-the-transmission efficiency the engine is doing almost as much work to propel the vehicle at 55 mph as it would have to do under more favorable conditions

For strong performance and economy, a Holley "Z Series" intake manifold is just about ideal. The manifold alone can be worth one or two mpg as compared to a factory dual-plate model.

Although intended primarily as a power-increasing option, a three two-barrel arrangement can work well in an economy application. The primary carburetor can be tuned to minimize gas consumption during cruise, while secondary carbs are jetted rich for maximum wide-open-throttle power.

In conjunction with the "Z Series" manifold, Holley recommends a Model 4160 four-barrel rated at 600cfm. This particular type of carburetor offers excellent throttle response in addition to good mileage.

For maximum economy, it is absolutely essential to use the correct rearend gear. Gears of about 2.7 to 3.1:1 ratio should be just about right for vehicles not equipped with an overdrive transmission.

EFFICIENCY	ENGINE (RPM) AT 55 MPH	SPEED (MPH) W/O SLIP
98	2245	56.1
90	2444	61
85	2588	65
80	2750	69
75	2933	74

to attain a 70 mph speed. (Even though engine speed is the same, wind resistance and frictional losses are less at 55 than they are at 70.) Depending upon the engine/chassis/body configuration of a vehicle, that extra engine speed can cost one to four miles per gallon. Overdrive transmissions and lock-up torque converters obviously make a significant contribution to fuel economy by reducing the engine speed/vehicle speed ratio dramatically.

Rear axle ratio also exerts considerable influence over fuel economy and should be chosen according to individual vehicle operating conditions. For the "typical" 3,000-3,500 pound car, the optimum ratio seems to be in the area of 2.7:1 to 3.1:1. Many late-model vehicles are fitted with axle ratios as high as 2.3:1, and while these are effective in minimizing engine speed, they frequently do not offer sufficient torque multiplication to allow maintenance of strong manifold vacuum. There have been cases where switching to a lower ratio (higher numerically) gear set actually improved fuel economy. This is especially appropriate with vehicles weighing in excess of 4,000 pounds.

Once the final combination of carburetor, intake manifold, exhaust system, ignition system and rear axle ratio is achieved, trial-and-error testing will usually provide the key to the maximum number of miles from a tankful of gas. Keeping the mixture slightly on the rich side will allow determination of optimum

Off-roading, whether competitive or casual, presents a number of unique fuel metering problems. Operating conditions vary greatly, and there's always plenty of dirt and dust around to cause problems.

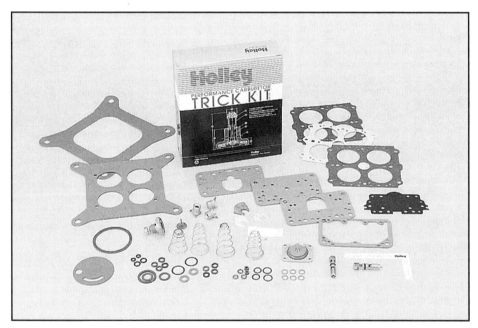

Considering the frequency with which off-road carbs must be cleaned or rebuilt, keeping a Holley Trick Kit around is a good idea. In addition to a typical assortment of rebuild gaskets, Trick Kits include extra parts for custom tuning.

ignition timing with minimal pre-ignition encounters.

With any knock and ping problems resolved, air/fuel ratios may be altered until the optimum setting is found. With experimenting going on, it's easy to lose sight of a few common mileage thieves:

1) Choke — may be opening too slowly or never reaching wide open. Bi-metal elements can become inoperative and hot air tubes can become clogged or disconnected.

2) Air Cleaners—restrict airflow when dirty, causing mixture to richen.

3) Clogged air bleeds—cause erratic operation and rich mixture.

4) Float level too high—contributes to an excessively rich mixture.

5) High idle speed — increases engine RPM and unnecessary gas consumption.

6) Inoperative heat riser valve—if stuck open, delays heating of intake manifold, which assists in fuel vaporization when engine is cold; if stuck closed, presents a restriction to exhaust gases.

7) Incorrect power valve—valves that open early (8.5-10.5 inches) may be providing enrichment before it is needed, causing an obvious waste of fuel.

As is the case with racing engines, there is no singular optimum "mileage motor" combination. What works for one driver may not be satisfactory for another. In fact, the same modifications applied to two different engine/chassis configurations may yield contradictory results, especially on late-model engines which are saddled with an infinite variety of emissions control equipment. Trial-and-error testing and a light throttle foot are still the best tools around for achieving maximum economy.

OFFROAD

Of all the various motorsports, off-roading is easily the most demanding on a carburetor. In no other motorized activity is there such a diversity of operating conditions; one minute a vehicle may be lugging along at 15 miles per hour, the next it is charging across a dry lake at more than 100 mph. Wildly fluctuating engine speeds and severe load variations are combined with impossible terrain and climate, making consistent fuel metering a difficult task.

The high speeds that are common in sanctioned off-road competition add to the difficulties, but even during a casual bash through the boonies, enough vibration, dust, dirt, slamming and banging are present to coax a stock carburetor into a nervous breakdown. Carb modifications are in order, yet they are surprisingly minor. Even for a full-on racer, only a few dollars worth of parts and a Saturday afternoon will turn a box-stock Holley into a backwoods superstar.

As a rule, only vacuum-secondary model Holleys should be used for rough-terrain off-roading. Mechanical actuation of the secondary throttle shaft opens the plates too quickly for many situations. The result is an excess of horsepower at the driving wheels, causing them to spin in loose sand, gravel, or mud. Rather than

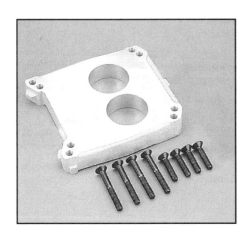

Holley offers a complete selection of carburetor adapters and, while it is preferable to use the proper manifold, an adapter will suffice if a manifold with a suitable Holley mounting flange is not available for a particular engine.

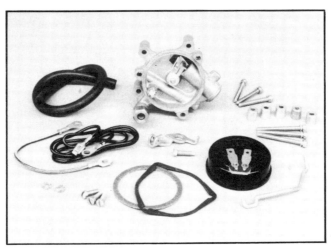

For special applications, the Holley integral electric choke conversion can be a lifesaver. This conversion uses a standard 12-volt electrical supply to heat an internal bi-metal coil that, in turn, controls the choke blade and fast-idle setting. This permits choke operation when other mechanical adaptations are not possible.

providing propulsion, the spinning wheels dig holes. Vacuum secondaries can be made to function more smoothly and are easier to control when the car and driver are bouncing all over the countryside.

Holley offers a number of 600cfm, Model 4160, carburetors for late-model applications. Selection of the part number designed for a specific vehicle type will alleviate linkage and mounting problems. The 600cfm capacity should provide an almost optimum combination of low-speed throttle response and high-speed horsepower. Quite obviously, larger displacement race-only engines require higher capacity carburetors to achieve their full horsepower potential.

Irrespective of carburetor model or size, when the going gets rough, the fuel level control in a standard float bowl can become a problem. As a truck bounces through the outback, its gyrations throw gasoline about the float chambers like a politician slinging mud. Frequently, the float will slam the inlet needle shut, even though the fuel level is low. If this occurs at an inappropriate time, a momentary leanout may result. The Holley off-road fuel bowl conversion kit (part no. 34-3) is designed to correct the shortcomings of standard bowls.

This kit will fit only Model 4150 and 4500 carbs, so if a 4160 is being used, it will have to be converted kit number 34-6) to the 4150 configuration. Improved fuel control is derived from the spring-loaded needle-and-seat which is used in conjunction with a spring-loaded float. In essence, this arrangement provides full suspension for the needle plunger, thereby damping erratic movement that is created by fuel being thrown about in the bowl.

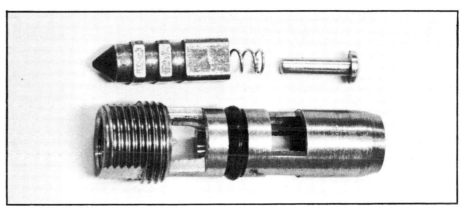

A special spring-loaded needle-and-seat is the key to stable inlet fuel control in rough-and-tumble off-road conditions. This assembly is included in the Holley off-road fuel bowl kit and is also available separately (part number 6-513).

The spring-loaded, needle-and-seat assembly (part number 6-513) may be installed in any side-inlet bowls, however, operation may not be totally satisfactory. Standard bowls use a weaker float bumper spring than the off-road units. Shimming the stock spring, or exchanging it for a stiffer one, may be required.

For most off-roaders, fuel control problems are not limited to the inlet circuit; just keeping fuel inside the bowl can be a problem. On Holley modular carburetors, the bowl vent is the major concern. The addition of a baffle inside the bowl and/or an extension of the vent tube is recommended. Either vent baffle 26-40 (commonly called a whistle) or vent screen 26-39 may be installed to minimize the amount of fuel that escapes the bowl. A rubber hose or a length of steel tube can also be used to increase the length of the stock vent tube. This will prevent any errant fuel from sloshing up the vent tube and reaching the venturi. Whatever means are used to extend the vent tube, the overall length should remain approximately 1/2-inch from the air cleaner top. The preferred extension method involves removal of the original tube and replacement of a longer version with the exposed end cut on an angle as on the original tube.

As much as anything else, periodic maintenance is essential to successful long term off-road carburetion. The preponderance of dirt and dust in this environment dictates that air and fuel filters be checked on a regular basis and changed when necessary. Whenever an inspection is made, the air bleeds should also be examined to insure that they are free of foreign matter. A quick squirt with Gumout or an equivalent spray cleaner will remove all but the most stubborn blockages. (Think of Gumout as the Drano of the carburetor world.) If this

Whenever an off-road spill-over is a possibility, a length of rubber hose or a longer-than-stock vent tube assists in containing fuel.

Although main jet slosh tubes are generally used for drag racing, they can be valuable as an addition to an off-road carburetor. Tubes keep the fuel from running away from the jets during hard acceleration or on inclines.

doesn't work, a very fine, soft wire may be used to make certain the bleeds are free of debris.

In most respects, the optimum off-road carburetor may be built using the same techniques used to tailor a Holley for street or race usage. Modification details may be found in other chapters of this book.

MARINE

For some reason, there seems to be an inexplicable communications gap between auto and boat enthusiasts. Carburetor problems that are totally foreign to land-based vehicles rise to the surface with disconcerting regularity whenever a boat hits the water.

To a large degree, marine carburetion anomalies arise from misapplication of carburetor/intake manifold configurations. The fact that a boat is rarely, if ever, driven in bumper-to-bumper traffic seems to infer to some owners that exotic racing induction systems are perfectly acceptable — irrespective of cam timing, compression ratio, or lack of internal modifications. Since bigger is obviously better, nothing less than a Model 4500 or two (at the least, a pair of 850 double-pumpers) is sufficient. Then spark plugs can be fouled on a regular basis and erratic engine operation can be assured.

If a 2x4 ram-style intake manifold on a mild small block is an absolute must for acceptance on the high seas or inland waterways (and the engine in question is not a race motor), use of two number 0-8007 Model 4160 carburetors should be considered. These carbs feature vacuum secondaries, electric chokes, side inlet float bowls and 390cfm airflow capacity. A pair of these on a Pro-Dominator intake manifold will provide good fuel distribution, and a total flow capacity of 780cfm. Such a configuration has been used successfully on several street rods, where "show" is as important as "go." More aggressive small blocks and big blocks of all descriptions will obviously benefit from a pair of 600cfm four barrels. Holley offers a variety of these with either vacuum or mechanically actuated secondaries.

One pitfall concerning use of two vacuum-secondary carburetors is a possible disparity in secondary throttle opening rate. This occurs when one carburetor opens earlier and receives a stronger signal than the other. The result is uneven fuel distribution. An easy remedy is to install a balance tube between the two vacuum diaphragm chambers. Ford used this arrangement in the '60s on dual-quad-equipped 427 engines. The simplest way to connect the chambers is to install the special vacuum diaphragm chamber tops (part no. 20-28) which contain a provision

A 454 Chevrolet with a single Holley is an ideal engine for a ski boat. A single carb makes tuning easy and provides low-speed driveability with strong top-end power, as proven on this dyno test.

Ram manifolds with a single carb may look impressive, but fuel distribution problems plague such installations. A single- or dual-plane manifold is a far better choice.

for connection of an external tube. If these are not locally available, the standard top may be removed, drilled and tapped, and replaced.

For competition-only race engines, a pair of 0-6109 (750cfm each) or 0-9375 (Model 4500's with annular discharge booster venturis, 1050cfm each) carbs will prove more than adequate. These carburetors may be tailored for maximum performance by following the guidelines presented in the chapters dealing with race modifications.

The pleasure boater who is more concerned with smoothness and consistency would be wise to use a single four-barrel on an appropriate performance manifold. By keeping carburetor size in the 750-800cfm range, low-speed "dockability" will be good, as will operating efficiency.

A number of years ago, Holley released a line of carburetors designed specifically for marine installations. These carburetors, which range in air flow capacity from 300 to 1250 cfm, meet Coast Guard requirements for fuel containment in the event of fuel flooding.

All Holley marine carbs feature J-shaped tube bowl vents which route any fuel escaping from the float bowl into the carburetor throat. The throttle shafts in marine carburetors are also unique in that they are grooved and slabbed to prevent fuel from leaking out of the carburetor.

The special construction of marine carburetors is inspired by the need to minimize the possibility of fire or explosion in the event a carburetor floods or the engine backfires. These concerns are primarily associated with pleasure boats in which the engine resides in a sealed compartment. Many ski and race boats have their engines positioned out in the open. Since fuel and fumes only accumulate in a closed area, the risks of fire and explosion are minimal. Consequently, automotive carburetors are commonly installed in these types of boats. However, this may be in violation of Coast Guard regulations, and a change to marine carburetors may be required.

Carburetor size and type notwithstanding, since boats are operated at wide-open throttle on a fairly regular basis, a standard mechanical fuel pump may not be capable of delivering adequate quantities of gasoline. Holley's marine mechanical and electric fuel pumps contain a fuel/fume vent tube and offer increased fuel delivery capacities. The marine VoluMAX electric pump is rated at 50 gallons per hour and the marine mechanical pumps squirt out either 110 or 130 gallons per hour. These capacities are more than ample for any single carb installation, provided fuel lines at least 3/8-inch in diameter and a high-capacity fuel filter are also installed.

Unless carburetors are mounted sideways, side inlet fuel bowls (rather than dual feed bowls) must be used for adequate clearance on dual-quad inductions. A special plate that fits below the carburetors provides linkage and fuel line mounting facilities for a clean-looking installation.

Even recreational boat owners heavily favor dual four-barrels, which make over-carburetion a regular occurrence on the water. A single four-barrel will provide equal mid-range performance. The additional airflow benefit of dual carbs is gained only at wide-open-throttle, near the maximum speed of the boat.

This 460 Ford engine is slated for installation in a day cruiser and uses a 780cfm vacuum secondary Holley. It will provide many hours of trouble-free operation.

Super Tuning and Modifying HOLLEY CARBURETORS
The Model 4360

Model 4360 carburetors were discontinued from production several years ago. They were originally available for a variety of GM and Chrysler V-8 engines and some Ford small blocks. Relatively little information about these carburetors is available, and considering that a fair number were produced, this chapter has been included.

Although a four-barrel carburetor is usually not thought of as being suited for installation on economy-type engines, both the Model 4150/4160 standard flange and the 4165 Spread-Bore designs have proven to be extremely versatile. Both have been used successfully in a variety of mileage-improving applications. However, the basic design of these carburetors was conceived a number of years prior to the days when the spotlight of governmental regulations was focused on exhaust emissions and gas mileage requirements.

Holley's Model 4360 was introduced in early 1976 and was the result of an intense research and development program that began in 1974. The 4360 was Holley's lowest cost Quadra-Jet replacement carburetor and it was furnished with all the essentials to make installation a simple bolt-on procedure. The 4360 construction is a radical departure from that of "standard" Holley modular carbs. Also unique was the first use by Holley of aluminum alloy for all major castings.

Holley engineers concentrated on low emissions and good fuel economy as primary design objectives, but the Model 4360 also qualified as a performance carburetor for smaller displacement engines (230-305 cubic inches). The relatively small primary throttle bores provide exceptionally good throttle response in the lower and middle RPM ranges, and the mechanically actuated secondary throttles provide excellent acceleration characteristics and good power output at higher RPM levels.

When installed on a medium-displacement engine (327-400 cubic inch small blocks), the Model 4360 provides excellent driveability and gas mileage with low exhaust emissions. Understandably, the comparatively low airflow capacity limits the maximum performance potential of a larger engine. It should also be noted that a Model 4360 may prove to be unsatisfactory when installed on a large-displacement engine with a radical camshaft. The small primary venturis are extremely sensitive to airflow and the pulses created by a long-duration camshaft may cause unsatisfactory fuel metering and atomization.

With 1 3/8-in. primary and 1 7/16-in. secondary throttle bores, the Model 4360 is rated at 450cfm when tested at wide-open throttle with a pressure drop of 1.5 inches of mercury (industry standard test criteria for four-barrel carburetors). However, since only the small primaries will be used for most driving situations, higher airflow velocities can provide a considerable increase in gas mileage over most two-barrel designs.

Although the 4360 uses mechanical actuation to open the secondary throttle plates, the accelerator pump system services only the primary

The Holley Model 4360 was introduced in 1976 as a low-cost replacement for the Rochester Quadra-Jet. A totally new, lightweight design, the 4360 provides economy and performance for small-displacement or low-horsepower engines.

Viewed from the underside, the simplicity of the 4360 becomes apparent. Secondaries are mechanically actuated and a myriad of vacuum sources allows for easy bolt-on installation on a wide variety of engines.

throttle bores. Theoretically such an approach should result in an objectionable stumble if the secondaries were opened rapidly. But the 4360 does not suffer from this malady. The small venturi size, coupled with an extremely long booster venturi, is responsible for a fuel signal that is very sensitive to changes in airflow and vacuum. This high sensitivity is one of the reasons for the 4360's incompatibility with radical cam timing.

The Model 4360 is comprised of three major die castings—the air horn, main body, and throttle body. These pieces are cast in high-quality aluminum alloy rather than the zinc alloy used in the Quadra-Jet and in other Holley designs. Aluminum is considerably lighter than zinc (a complete 4360 weighs a mere 6.4 pounds; a Quadra Jet tips the scales at 8.8 pounds) and is very stable at high temperatures, providing improved long-term gasket sealing. Holley also employs a special resin-coated gasket at the main body/throttle body and main body/air horn junctions. When normal underhood temperatures are applied, a catalytic reaction between the gasket and gasket surfaces creates a virtually leak-proof seal that is extremely

The Model 4360 is ideal for use on carbureted Buick V-6 engines.

durable. This exceptionally strong bonding of gasket-to-surface can create disassembly problems. When attempting to disassemble a 4360 (as for modification or rebuilding), separating the air horn from the main body can be a sizable chore. Frequently, the gasket will tear before it pulls loose from either carburetor surface. Efforts to loosen the gasket should be undertaken carefully, since vigorous prying with a screwdriver or similar tool can damage the gasket mating surfaces. Prying gently and "working" the gasketed components is the most effective means of separating the desired parts.

The main body is where most of the "carbureting" takes place, and the other major components (air horn and throttle body) are basically supportive to the actions which occur in the main body. The overall simplicity of this design makes the 4360 very easy to understand and troubleshoot. The best way to gain insight into the operation of the 4360 is to follow the flow of air and fuel through the carburetor.

BASIC FUNCTIONING

Gasoline enters through a single .110-inch diameter inlet needle-and-seat (part no. 6-514) in the main body and must flow through a standard Holley sintered bronze filter before reaching the float chamber. The design of the float chamber comprises a radical departure from former Holley practices. Rather than being located at the end of the carburetor, the 4360 chamber is a long, narrow affair which runs between the left and right throttle-bore pairs. The chamber will hold 90cc of gasoline and is deeper at the rear, thereby preventing fuel starvation during acceleration. Float chamber volume and jet location is such that starvation is not encountered during decel-

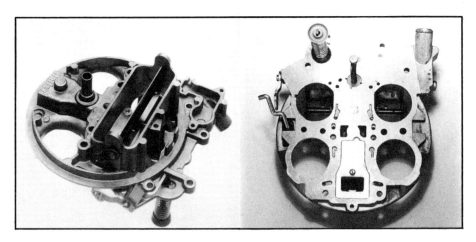

The upper casting contains the choke plate, accelerator pump plunger, and power valve operating mechanism. It also serves as a cover for the float bowl, and it is easily removed to gain access to the internal parts.

Beneath the metal plate in the upper casting is nothing more than a vent chamber. The plate is part of the baffle arrangement that prevents fuel from leaking out the vent tube.

eration, nor is it a problem during severe cornering. Unless the carburetor is physically turned upside down, it is almost impossible for gasoline to evade the main jets.

After flowing through the main jets, fuel enters the main well and passes into either the idle or main metering fuel circuit. During engine idle, low pressure is presented at the curb-idle discharge port and fuel is drawn out of the idle well. The idle well is fed fuel from the main well through the idle-feed restriction. The fuel from the idle well travels across one of the four tracks (one for each venturi) machined in the top of the main body, where bleed air is introduced through a channel restriction and finally down to the discharge port and transfer slot. The track (on top of the main body) over which the air/fuel mix travels is clearly visible when the air horn is removed. Throughout the carburetor, restrictions in the idle circuit consist of brass bushings which may be modified for non-standard applications.

Regulation of the idle mixture is accomplished through use of a standard — clockwise/lean, counterclockwise/rich — mixture screw. Although idle mixture screws are included only on the primary side, it should be noted that the 4360, like all four-barrel carbs, has a non-adjustable idle system which feeds the secondary throttle bores.

Located above each throttle plate is a transfer slot which comes into play during the transition from idle to open throttle. Since there is always at least a slight opening between the throttle plate and body, a small amount of air/fuel mix is discharged through the transfer slot even during engine idle. When the throttle plate begins to open, it uncovers more of the transfer slot, exposing it to manifold vacuum, causing a higher flow (of air/fuel mixture) through the slot than existed during idle. Once the main metering circuit is activated, the metering signals shift and the idle circuit and transfer slot fade from importance. The secondary idle circuit is identical to the primary side except for the lack of mixture adjustment and idle discharge. All secondary idle fuel is discharged through the transfer slot, except on carburetors intended for V-6 applications. (A complete description of typical idle circuit operation is contained in Chapter 4.)

As the throttle is opened further and airflow increased, venturi vacuum increases correspondingly and the relative low pressure produced in the venturi area draws fuel from the main well. Once past the main well tube it is mixed (emulsified) with air admitted through the high-speed bleed and enters the air stream through the discharge nozzle in the booster venturi. At this point, the main metering system is handling virtually all the fuel flowing through the carburetor.

During normal cruise, manifold vacuum is relatively high and the engine fuel requirements are stable. The main metering jets are selected primarily to handle low-load cruise fuel requirements and will deliver mixtures that tend to be too lean for high-load conditions such as those present during acceleration or when climbing a steep hill. Therefore, two additional circuits are employed to provide increased operational flexibility. They are the power enrichment system, which allows extra fuel to flow into the main well when manifold vacuum drops below a predetermined point, and the accelerator pump system, which squirts fuel directly into the primary venturis each time the throttle is quickly opened. Each of these systems is designed to accommodate different driveability requirements.

As with most carburetor designs, the 4360 uses an accelerator pump circuit to eliminate the stumble that typically accompanies rapid opening of the throttle. Other Holley models use a cam-actuated lever/diaphragm pump to move fuel from the pump well to the discharge nozzle. The 4360 departs from this Holley tradition, using a rod-and-lever actuated, spring-loaded plunger to perform the pumping function. The pump plunger is driven down by spring pressure and returned to the cocked position by the linkage when the throttle is returned to idle. Fuel is admitted to the pump well through clearances built into the center of the plunger cup. Refilling occurs when fuel flows through these openings as the cup raises during throttle closure. When the throttle is opened, the downward movement of the cup closes the clearances and forces fuel past the outlet check ball and through the discharge nozzles. Kill bleeds, which are little more than passages leading to a higher-than-venturi-pressure air source, are above the pump-discharge nozzles to prevent incoming high-velocity air from siphoning fuel out of the fuel supply passages located below the nozzles.

The power enrichment circuit is not like the 4150, but is similar to other Holley OEM one- and two-barrel models. It is a two-stage design. Manifold vacuum actuates a spring-loaded piston-and-rod which, in turn, operates the power valve needle. During periods of high manifold vacuum (cruise and idle) the piston is drawn up, allowing the valve closing spring to push a tapered needle valve against the seat. When manifold vacuum drops, the power valve operating spring forces the piston-and-rod down, overcoming the pressure exerted by the valve closing spring, and opening the power valve. This admits additional fuel through the channel restrictions to the primary main wells.

The first models of the 4360 were calibrated for passenger car engines displacing 327-455 cubic inches. Later models were added to accommodate V-6 and small (260-305) cubic inch V-8 applications. Mixture calibrations differ in areas of jet size, power valve opening point, idle mixture, secondary idle circuit, and accelerator pump capacity and duration.

Part numbers 0-8156 (intended for the 1975 through 1977-1/2 and later V-6s), 0-8517 (calibrated for the 262 and 305 Chevrolet V-8s) and 0-8479 (designed for the Olds V-8) differ significantly from models which are intended for larger V-8 powerplants. The original Buick V-6 odd-firing configuration created pulsing in the intake manifold, apparently causing the air/fuel ratio to become richer. Whereas the carburetors designed for the later even-fire V-6 and small V-8s use no. 203 (.047-inch) or no. 219 (.048-inch) main jets (part no. 124-XXX, the last three numbers refer to jet number), the odd-fire carb had to be jetted considerably leaner with no. 167 (.044-inch) jets. On the dyno the V-6 ran well with no. 139 (.040-inch) jets but the richer no. 167s were required for acceptable driveability. As with all Holley jets the number refers to flow rating, not diameter.

The V-6 is also very sensitive to off-idle throttle openings and the introduction of exhaust gases through the EGR (exhaust-gas recirculation) valve. In order to improve fuel economy and driveability, both the 0-8516 and the 0-8677 use .040-inch wide transfer slots rather than the .025-inch wide slots used in other models.

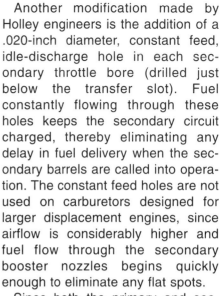

The main body contains most of the fuel metering mechanisms, including the float chamber, booster venturis, jets, power valve, and the fuel delivery passages.

This view of the main body illustrates why the float is relatively insensitive to side-to-side movement (cornering). The long, thin float chamber doesn't allow the fuel much sideways movement. Conversely, radical fore-and-aft movement (acceleration, braking) significantly can affect fuel level.

Another modification made by Holley engineers is the addition of a .020-inch diameter, constant feed, idle-discharge hole in each secondary throttle bore (drilled just below the transfer slot). Fuel constantly flowing through these holes keeps the secondary circuit charged, thereby eliminating any delay in fuel delivery when the secondary barrels are called into operation. The constant feed holes are not used on carburetors designed for larger displacement engines, since airflow is considerably higher and fuel flow through the secondary booster nozzles begins quickly enough to eliminate any flat spots.

Since both the primary and secondary systems pull fuel from the same bowl, lack of constant feed idle discharge holes in the secondary throttle bores will not cause stale fuel problems, as would be experienced in a carb with a separate secondary fuel bowl. Other factors are carburetor size and the inclusion of mechanical secondaries. With the small primary bores of the 4360 it is virtually impossible to keep the secondaries closed for extended durations. And, since secondary actuation is

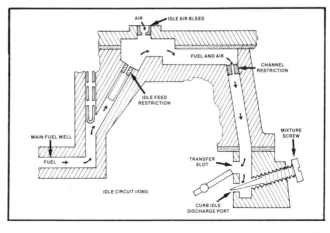

Operation of the 4360 idle circuit is typical of all modern carburetors. Fuel is drawn from the main well, bleed air is induced to provide an emulsified mixture, and the fuel emulsion is introduced into the airstream immediately below the throttle blade.

Super Tuning and Modifying Holley Carburetors **109**

Looking down the throat of a 4360, placement of the air bleeds and booster venturi configuration is visible. Secondary boosters are extremely sensitive to changes in airflow.

The extremely long booster venturi is visible in this view of the primary bore. Booster design is sensitive to changes in airflow.

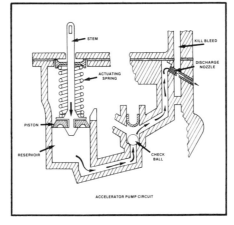

Instead of the diaphragm-type pump found on the traditional Holley performance carbs, the 4360 uses a piston-type accelerator pump. The pump plunger is forced down by sudden throttle opening. Fuel is pushed out of the reservoir into the delivery passages to the discharge nozzle. The check ball keeps the delivery circuit primed for immediate fuel delivery.

The power valve resides in the bottom of the float chamber, near the front. The needle in the center of the valve controls the flow of fuel. It is pushed open when the manifold vacuum drops.

mechanical, a slight bit of overenthusiasm will open the secondary throttle plates, causing secondary fuel flow.

In a small V-8 there may be problems with relatively low airflow signals delaying activation of the secondary booster discharge nozzles. This problem is created by the longer manifold runner passage length required for a V-8 engine. In order to eliminate momentary lean conditions when the carburetor is rapidly brought to wide-open throttle, part number 0-8517 is equipped with a larger capacity accelerator pump and a longer pump stroke, which continues fuel discharge during the initial stages of secondary throttle opening.

MODIFICATIONS

Since the 4360 is not of the same modular design as standard Holley four-barrels, methods of modification are not quite as easy. A variety of jets and two-stage power valves are available, but they are about the only components that can be easily removed and replaced. All other components must be individually modified for improved performance.

POWER VALVE

Power valve opening points can be altered by modifying the effective pressure of the power valve main spring. Placing a shim (or shims) between the spring and power valve piston retainer (which is staked in the air horn) will increase spring pressure, causing the power valve to open sooner (i.e., at higher manifold vacuum). Cutting the spring reduces tension, delaying power valve opening until manifold vacuum drops to a lower point.

However, note that most Model 4360s are fitted with a two-stage power valve. If the operating piston spring is altered, the timing of the first and second stages will be changed by the same amount. For example, if the operating spring tension is reduced to gain earlier opening of the first stage, the second stage will also open earlier (by approximately the same amount). Currently, there is little specific information concerning the 4360 power valve assemblies (there are several different ones), so modifications will have to be checked by trial and error.

ACCELERATOR PUMP

Since the Model 4360 uses a rod-and-lever mechanism to effect accelerator pump operation, the only means of modifying pump operation is to bend the pump actuator rod. Shortening the rod serves to raise the static height of the pump plunger, which increases available fuel volume in the accelerator pump well. The rod should be bent carefully and checked to verify that it is not binding at any point in its travel. Care should also be taken to insure that the rod is not altered an excessive amount. Should this occur, the pump plunger cup may be pulled entirely out of the well, causing damage to the cup lip or complete loss of the accelerator pump circuit. In all cases the rod should be located in the original, uppermost hole in the pump operating lever. A heavier pump shot can also be achieved by enlarging the discharge orifices. It can be drilled as large as .035-inch in diameter (as opposed to the standard .028-inch diameter). Judicious use of the drill held in a pin vise is advised, since opening of the nozzle excessively may cause a loss of performance.

FLOAT LEVEL

Before performing any modifications to the accelerator pump circuit it is advisable to check fuel level in the float chamber. Maintenance of proper fuel level is essential to satisfactory accelerator pump operation. The cylindrical wall which surrounds the accelerator pump plunger has a spillover ledge approximately 3/8-inch below the normal operating fuel level. A low fuel level setting can delay refilling of the pump well. Checked with the air horn assembly removed, fuel level should be set at 5/8-inch (+1/32-inch) below the top of the main body metal surface (as opposed to the gasket surface). If fuel level is considerably below this point, incomplete filling of the pump will result and off-idle performance will be accompanied by hesitation. Inadequate fuel volume will also cause a stumble when the secondaries are opened because the pump circuit will not be fully charged.

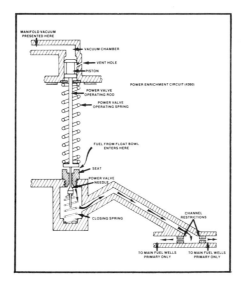

Vacuum activation of the 4360 power enrichment circuit is similar to most existing designs, but the hardware is unique to the 4360 model. During periods of high vacuum (cruise), the power valve needle is pulled up against the seat, closing off fuel delivery to the power circuit. As vacuum drops off, the operating spring pushes the needle away from the seat and fuel from the float bowl enters the circuit for delivery to the main fuel wells.

Believe it or not, this is a Holley power valve. The 4360 utilized a totally different design than the 4150/4160-type carburetor. Optional power valves are not generally available, but the opening rate can be changed somewhat by modifying the operating spring pressure.

The power valve is controlled by a spring-loaded rod in the air horn body. The rod is actuated by a vacuum-controlled piston. When the manifold vacuum drops to a predetermined level, the spring overcomes the vacuum applied to the piston, pushing the rod downward, and opening the power valve needle.

Three holes are contained in the accelerator pump operating lever; however, only the uppermost hole should be used. Accelerator pump operation may be modified by bending the link rod.

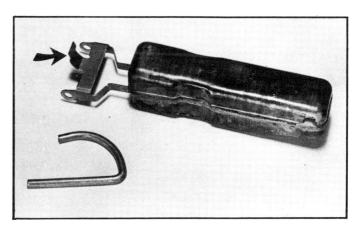

The fuel level in the float bowl is not externally adjustable as on the 4150-4160 model. The fuel bowl level is altered by removing the top casting and bending the tab that contacts the inlet needle valve.

Super Tuning and Modifying Holley Carburetors

REBUILDING AND TUNING THE MODEL 4150/4160

Rebuilding a Holley Model 4150/4160/4165/4175/2300 carburetor is an easy job. However, if you've never done it before, it is a good idea to read through this section to familiarize yourself with the various components and procedures. The example shown here is a typical late-model 4160 with vacuum secondaries. There may be some slight variations from the procedures shown here when rebuilding other Holley models, but the general methods are similar. The Holley Carburetor Company supplies illustrated instructions with most rebuild kits. These instructions should be consulted for specific details. The average rebuild can be completed with ordinary hand tools, however, it is recommended that the rebuilder have access to a compressed-air source to completely dry those components which are dipped in carburetor cleaner. It is always important to remember that gaskets, floats and plastic or rubber items should never be placed in carburetor cleaning solution. Before placing any component or assembly into caustic cleaner, make certain all pieces that may be damaged by the cleaner have been removed.

The disassembly operation will be considerably easier if the carburetor is properly supported during the operation. Use of a carburetor stand is advisable, but a short piece of wood placed under the throttle body may provide a suitable substitute. The intention is to raise the carb high enough to prevent damage to levers and linkages that protrude below the throttle body.

The first stop is to remove the primary float bowl. Four screws, located at each corner of the bowl, hold the entire bowl assembly in place. When the lower screws are loosened, residual gasoline in the float bowl will drain from the screw holes. This gas can be collected in a small cup or, before removing the bowl screws, it is possible to turn the carb bottomside up to allow the residual gas to drain through the bowl vents.

Once the carburetor has been in service for a long time, the bowl screw gaskets tend to stick tightly to the bowl. They can generally be reused, so they should be pried loose, and retained for future use. Replacement bowl screw gaskets are generally included in Holley rebuild kits, but the old gaskets will make suitable spares.

Newer Holley models are assembled with composition gaskets, rather than the traditional cork-type gaskets. Some of these composition gaskets are coated with a sealing resin that makes disassembly somewhat difficult. It may be necessary to gently pry the fuel bowl away from the metering block with a wide-blade screwdriver.

Once loosened, the fuel bowl can be easily pulled away from the primary metering block. The fuel transfer tube joining the primary bowl to the secondary bowl can also be removed. The O-ring seals on each end of the transfer tube should pull free along with the tube. These O-ring seals look as if they can be reused, but don't reuse them! They tend to expand after the first installation, and once they are removed it is very difficult (impossible!) to reseal them properly in the fuel transfer tube seat. This is not applicable, of course, on carbs with dual feed bowls because they do not use a transfer tube to supply fuel to the secondary bowl.

112 *Super Tuning and Modifying Holley Carburetors*

Once the float bowl has been removed, if you are planning to soak it in cleaning solution, you must complete disassembly of the bowl. The fuel inlet fitting may be removed with an 11/16-inch open-end wrench. Fittings on dual feed bowls require a 1-inch wrench. The bronze inlet filter, the inlet filter gasket, the inlet filter spring and the inlet fitting gasket can now be removed. The gaskets and the filter should be replaced but the filter spring can be reused.

The needle-and-seat assembly is removed next. First, loosen and remove the lock screw with a wide-blade screwdriver.

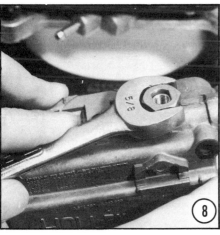

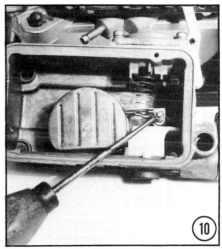

Then, turn the adjusting nut with a 5/8-inch wrench to pull the needle-and-seat assembly upward out of the bowl. Note that the needle-and-seat assembly is threaded into the bowl casting and that the adjusting nut pilots onto the top of the assembly.

When the needle-and-seat threads are clear of the bowl threads, the entire assembly can be pulled out of the bowl, but the O-ring seal around the body of the assembly may provide some resistance. Just pull straight upward, firmly. This procedure is the same for all externally adjustable Holley float bowls.

The float can be removed next. Once the C-slip is pulled free from the pivot shaft, the float can be pulled out of the bowl.

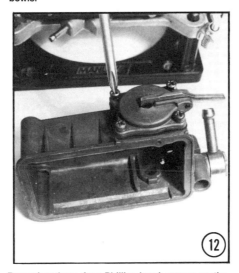
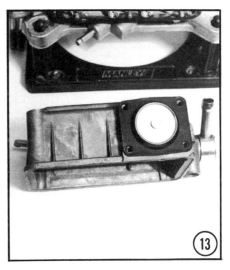

Also remove the plastic inlet baffle. This baffle is essential for fuel-level control and must later be reinstalled into the bowl. Fuel flow will not be increased by leaving it out. On dual feed bowls the inlet cavity serves as a built-in baffle, so this plastic baffle is not present.

Removing these four Phillips-head screws on the bottom of the primary bowl allows the accelerator pump cover to be lifted (or gently pried) from the pump housing.

Below the pump cover is the diaphragm. Replacement accelerator pump diaphragms are not included in all kits, so this piece should be handled carefully to prevent damage.

Super Tuning and Modifying Holley Carburetors

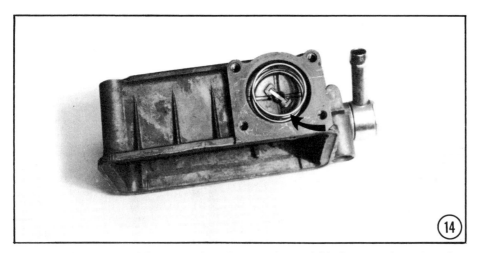

With the diaphragm removed, the pump cavity and return spring are visible. Remove and save the spring.

The pump inlet check ball is retained in the top of the pump cavity by this strap. Clearance between the ball and the retaining strap controls the refill rate of the pump and the discharge delay (the ball must be seated against the inlet opening before fuel is discharged at the pump nozzle). In 99.99% of the cases if the strap is left alone, pump performance will be entirely satisfactory. However, it is possible to check the clearance to verify that it is .010-inch with the bowl inverted.

A resin-coated composition gasket may be installed between the motoring block and the main body. Once again, a wide-blade screwdriver may be required to gently pry the block free.

On occasion, the gasket will tear, leaving material on the main body and/or the metering block surface. This material must be carefully scraped from all surfaces (this is easier after the parts are dipped in cleaning solution).

Many late-model Holleys use an accelerator pump transfer tube sealed at each end with an O-ring. The tube does not have to be dipped into the cleaning solution, and since removal of the O-rings is troublesome, it is generally possible to reuse the tube and O-rings.

The power valve can be removed from the block with a 1-inch socket, open-end or box. Care must be taken to avoid damaging the block surface.

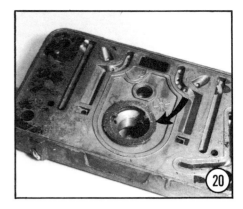

When the power valve is removed, the gasket seated between the power valve and the block should be chocked to insure that the new replacement gasket is of the same configuration. The gasket shown here is intended to be used with two-stage, four- and six-hole valves. The new picture-window-type valve requires use of a gasket without internal locating tabs. The correct type gasket must be used with each valve to insure proper power valve operation!

The idle mixture screws must also be removed from the sides of the metering block. On some models it may be necessary to remove the plastic idle-limit caps to gain access to the mixture screws. There is a small cork gasket that seals around the mixture screws; this should be pulled out of the recess in the block. Also, if there is an idle-indicator decal on the block, it should be removed before the block goes into the cleaning solution and reapplied during the assembly phase. This decal is present only on reverse-idle models.

Once the primary side is disassembled, the same general procedure should be applied to the secondary float bowl/metering assembly. On 4150 models the secondary float bowl and metering block are virtually identical in appearance to their primary counterparts. However, on Model 4160 carburetors, such as the one shown here, a special screwdriver is necessary to remove the clutch-head screws used to retain the metering plate to the main body. Note also, there will not be an accelerator pump in the secondary bowl-except on double-pumper models, all of which use a secondary accelerator pump assembly virtually identical to the primary pump.

The next step is to begin stripping the main body. The best place to start is to remove the choke assembly. The choke shown here is a Holley integral electric choke. This disassembly is typical of most integral-type chokes. Most OEM-type divorced choke assemblies can be removed from the main body as a unit, once the main operating rod (connected to the remote sensor or bimetal on the intake manifold) and choke-plate operating rod are unfastened from the choke assembly. Nearly all choke assemblies utilize plastic components, so they must be removed before the main body can be dipped in cleaning solution.

To remove the main choke housing, loosen the three Phillips-head screws holding the housing to the main body and remove the R-fastener from the lower end of the choke-plate operating rod. With the housing free, the choke-plate operating rod and the small plastic guide can be removed from the base of the choke shroud.

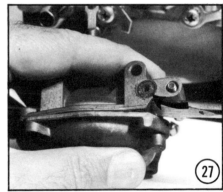

First, note the alignment of the choke adjustment index (this will allow you to reset the choke during reassembly) and then remove the three screws and the retaining ring holding the plastic choke cover in place. When you pull the choke cover free, note that the operating tab inside the main choke housing fits inside the ring on the end of the bimetal coil housed in the choke cover. When reassembling the choke, the operating tab must be inserted back into the ring. Some older carburetors have an L-shaped extension rather than a ring on the bi-metal, but choke operation is similar, therefore, the tab and the bi-metal extension must be aligned so that the choke blade is pulled close when the bi-metal is cold.

Similarly, three screws and a C-clip must be removed in order to separate the secondary diaphragm housing from the main body. (This does not apply to double-pumpers which use a mechanical secondary linkage.)

To further disassemble the diaphragm after it is free from the main body, remove the four screws that retain the diaphragm housing cover and the cover can be gently pried loose. Remove the diaphragm and the diaphragm return spring. Save both the diaphragm and the return spring unless you plan to replace them during the rebuild. Note the cork gasket around the secondary vacuum passage (above and to the left of the screwdriver). This gasket should be replaced during any rebuild and inspected whenever the diaphragm is removed. Should it become damaged or excessively compressed, it will not maintain a suitable seal, and secondary activation may be delayed or prevented altogether.

Super Tuning and Modifying Holley Carburetors

The final disassembly stop is to turn the carb bottom side up and remove the throttle body assembly from the main body casting. Usually six (or eight) Phillips-head screws hold the throttle body in place. Once the two castings are separated, pull the main body gasket off carefully and set it aside. Several different main body gaskets are usually contained in each rebuild kit (to suit different models), so it will be necessary to match the new with the old to insure proper selection. Remove the plastic accelerator pump cam from the inside of the throttle lever (do the same for the secondary pump cam if you're working on a double-pumper), and soak the principal components in carb cleaner for a few hours.

Once all the parts have been soaked in cleaner, washed and blown dry with high-pressure air, it's a good idea to group the components and gaskets for easy subassembly (a typical float bowl grouping is shown here). The general reassembly procedure will be the reverse of the disassembly sequence.

With everything else removed from the main body, the accelerator pump nozzle is next. Once the retaining screw is removed, the carb may be inverted with your hand held over the primary bores, and the nozzle and check valve should drop free. There is a thin gasket seated between the nozzle and the main body. Often this gasket sticks to the nozzle seat in the main body, so it may have to be scraped free. If you are rebuilding a double-pumper, there is also a pump nozzle on the secondary side. It is removed in a similar manner.

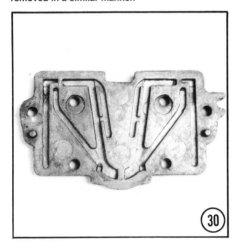

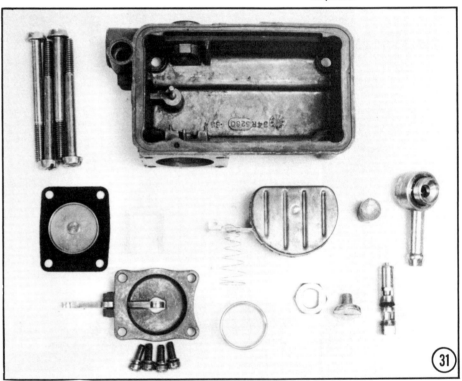

Since the internal passages are now completely exposed, the main body, the throttle body and the metering block(s) must be carefully inspected to insure that wayward pieces of gasket or other foreign material have not accidentally blocked a passage or metering orifice.

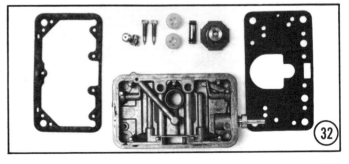

Before reinstalling the primary metering block, the main jets, the power valve, the idle-mixture screws and the accelerator pump transfer tube (if so equipped) should be returned to their respective places along with the appropriate gaskets. The fuel bowl and metering block gaskets should be carefully positioned during the installation of the bowl.

When reinstalling the secondary metering plate, first the contoured gasket is placed on the plate surface, then the thin metal plate is placed on top of the contoured gasket, and these three pieces are placed on the large rectangular gasket. Finally, the four-piece "sandwich" is positioned against the main body. The six retaining screws are installed and evenly tightened.

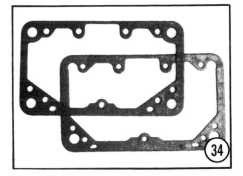

Caution: There are two types of fuel-bowl gaskets. Although similar in appearance, they are not interchangeable. The gasket at right is for standard Model 4150, 4160, 4500 and 2300 carburetors. To the left is a similar gasket for the Model 4165, 4175 and the three 4150/4160 part numbers that use Spread-Bore-style metering blocks (0-6708, 0-6709 and 0-7010).

Once the float bowl has been reassembled, a dry float level setting should be made to minimize start-up difficulties. With the bowl inverted, the flat (top) surface of the float should be parallel to the roof of the bowl.

This setting would result in a fuel level that is too high.

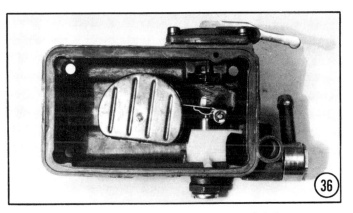

This setting would result in a low fuel level.

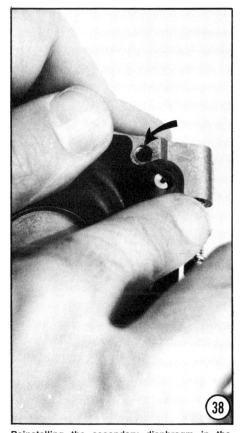

Reinstalling the secondary diaphragm in the diaphragm housing is very important, but it can be troublesome. There is an easy way. First, make sure that the diaphragm isn't torn and the seal around the housing vacuum passage is in place and in good shape.

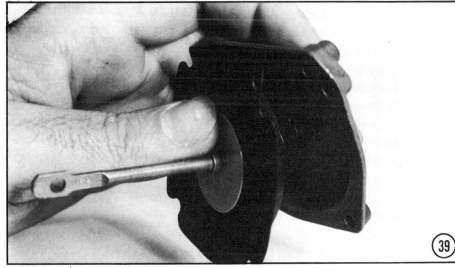

Then, with the diaphragm spring in place between the diaphragm and the housing cover, push the metal portion of the diaphragm toward the cover. Continue pressure on the diaphragm until the outer edges are seated against the cover. Inspect the edges of the diaphragm to verify that the screw slots and the vacuum transfer opening align properly (this is an important preliminary check).

Now, place the housing cover and diaphragm on a firm surface, bottomside up, and slide the diaphragm housing over the operating rod. Rest the diaphragm against the cover as in the previous step. The housing should seat against the edges of the diaphragm, sandwiching it squarely between the cover and the housing. Check all four sides to insure that the diaphragm hasn't slipped out of position. By holding pressure against the operating rod, the top can be lifted slightly and realigned, if necessary. Once the assembly is properly positioned, squeeze the cover and housing tightly and turn it over to insert the retaining screws.

Super Tuning and Modifying Holley Carburetors

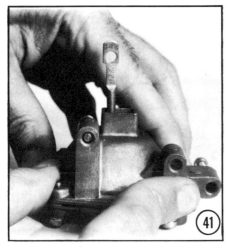

41 As a final check, after the screws are tightened, push the operating rod toward the housing top. Place your finger against the vacuum passage on the side of the housing, and release pressure on the operating rod. The diaphragm should not drop downward a significant amount. If it does, there is a leak. Check the diaphragm again (or check your finger).

42 This gasket must be seated properly or the Holley integral-type electric choke will not function correctly. Manifold vacuum pulls air through the choke housing to move heat from the internal heating element to the bi-metal coil. If the gasket is damaged, the choke opening will be delayed.

43 Needle-nose pliers will facilitate insertion of the accelerator pump check valve into the discharge passage. It should be dropped in first and ...

45 With the O-ring positioned on the very end of the fuel transfer tube, the tube may be gently pushed into the tube seat in the secondary fuel bowl. Slowly rotate the tube as it is inserted.

44 Then, the nozzle and screw may inserted in the same way, although the nozzle may require some maneuvering to get it into position below the choke plate. Two gaskets are used, one between the nozzle and main body and the other between the screw and the nozzle. Don't forget either one.

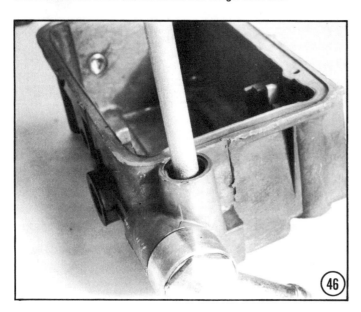

46 When properly seated, the O-ring ridge on the tube is inside the transfer tube seat opening.

47 With the primary bowl already mounted on the main body, the secondary bowl with the transfer tube properly seated in place may be maneuvered into position. To expedite the operation, the primary-side O-ring should be placed on the very end of the transfer tube, which is then guided into the primary tube seat. As the secondary float bowl assembly is pushed toward the main body, the tube should be pushed firmly into the seat. It sounds complicated, but it's easy when you get the hang of it.

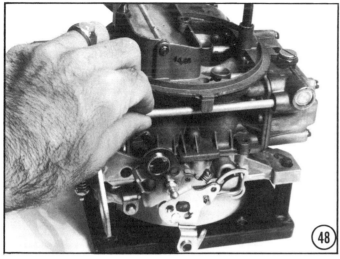

Once the secondary bowl is fastened in place, the transfer tube should rotate with some slight resistance. If it doesn't, the O-rings have not seated properly. Try again! Note that it is usually easier to mount the primary bowl to the main body first and maneuver the tube into place with the secondary bowl assembly. This way you don't have to contend with the primary accelerator pump linkage and the primary metering block when you're trying to get the tube and the bowl into position. If the carb has a secondary metering block (instead of a plate), you might as well start the assembly from either end.

Now that the carburetor is back together, a few adjustments should be made before reinstallation on the engine. In order to get the idle mixture screws "in the ballpark," turn them all the way into the seat (do not force them) and then back them out 1-1/2 full turns. Once the carburetor is actually metering fuel to the engine after startup and the float level has been checked, a final adjustment can be made. On a standard-idle system, with the engine running at the desired idle speed, each screw should be slowly turned clockwise (lean) until the idle rpm drops noticeably. (This verifies that the idle circuit for this particular bore is working properly.) Then, slowly back the screw out until the maximum idle rpm is reached. Repeat this procedure on the other idle mixture screw. Check to make certain that both mixture screws are approximately the same number of turns off the seats. This is important to verify that both the right and left portions of the idle system are operating correctly. Reset the idle-speed screw to the desired idle speed.

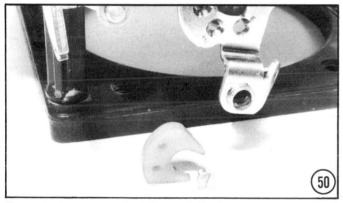

The accelerator pump cam should be mounted back in the original position and the pump operating lever should be readjusted. This is especially critical if a different pump cam has been installed or if the cam is installed in the second (optional) mounting hole.

This screw controls the idle speed when the choke is closed. This is called the "fast idle" speed and specific settings are usually given in the instruction sheets supplied with Holley rebuild kits. If the initial setting is not correct after the carb is installed and the engine is running, it may be necessary to remove the carburetor to readjust the setting. Holley uses various choke mechanisms and, fortunately, many fast-idle adjustments can be made with the carburetor in place. Nonetheless, you should check this setting before installing the carb.

With the throttle completely closed, the pump operating lever should be adjusted so that there is no clearance between the pump housing lever and the override spring assembly on the end of the operating lever. Adjustment may be made by holding the screw and turning the nut. Clockwise adds clearance, counterclockwise removes it. Also, check the pump override clearance. Push the throttle to the wide-open position and verify that there is an additional .015- to .020-inch clearance before the pump housing lever bottoms the pump diaphragm in the pump housing.

With the choke properly reassembled, the choke cover should be rotated until the index mark aligns with the desired ridge on the choke housing. The choke housing should be closed (cold) by pressure from the bimetal. Once final adjustment is made, all three retaining screws should be tightened. Final adjustment to fine-tune the choke operation may also be necessary after the carburetor is reinstalled.

After the engine is started, the float level should be checked and, if necessary, readjusted so that fuel just trickles out of the inspection hole. The screwdriver controls the lock screw, while the wrench turns the adjusting nut. The two must be turned simultaneously. Clockwise lowers the fuel level and counterclockwise raises the level. Loosen the lock screw only a small amount, otherwise fuel is likely to squirt out around the adjusting nut.

Super Tuning and Modifying Holley Carburetors

EXPLODED VIEW - MODEL 4500

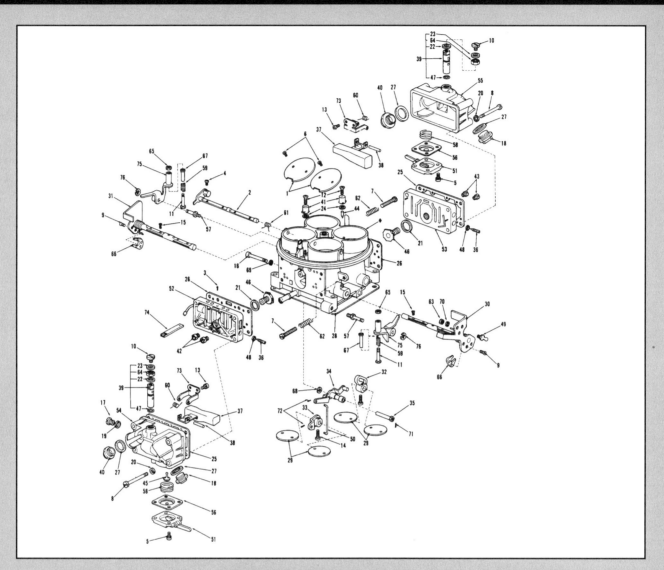

1. Choke Plate
2. Choke Shaft Assy.
3. Fuel Bowl Vent Baffle Drive Screw
4. Choke Shaft Swivel Screw
5. Fuel Pump Cover Screw & L.W.
6. Choke Plate Screw & L.W.
7. Throttle Stop Screw
8. Fuel Bowl Screw
9. Pump Cam Lock Screw
10. Fuel Valve Seat Lock Screw
11. Pump Operating Adj. Screw
12. Pump Discharge Nozzle Screw
13. Float Shaft Bracket Screw & L.W.
14. Throttle Shaft Screw
15. Throttle Plate Screw
16. Pivot Screw
17. Fuel Level Check Plug
18. Fuel Inlet Plug
19. Fuel Level Check Plug Gasket
20. Fuel Bowl Screw Gasket
21. Power Valve Gasket
22. Fuel Valve Seat Adj. Nut Gasket
23. Fuel Valve Seat Lock Screw Gasket
24. Pump Discharge Nozzle Gasket
25. Fuel Bowl Gasket
26. Metering Body Gasket - Pri. & Sec.
27. Fuel Inlet Fitting & Plug Gasket
28. Flange Gasket
29. Throttle Plate
30. Throttle Shaft Assy.-Pri.
31. Throttle Shaft Assy.-Sec.
32. Primary Throttle Lever (internal)
33. Secondary Throttle Lever & Bushing Assy.
34. Intermediate Throttle Lever Assy. (Comp.)
35. Threaded Guide Bushing
36. Idle Adjusting Needle
37. Float & Hinge Assy.
38. Float Shaft
39. Fuel Valve & Seat Assy.
40. Fuel Inlet Fitting
41. Pump Discharge Nozzle
42. Main Jet-Pri.
43. Main Jet-Sec.
44. Pump Discharge Needle Valve
45. Pump Check Valve
46. Power Valve Assy.
47. Fuel Valve Seat "O" Ring Seal
48. Idle Adjusting Needle Seal
49. Throttle Lever Ball
50. Connecting Link
51. Fuel Pump Cover Assy.
52. Metering Body & Plugs Assy.-Pri.
53. Metering Body & Plugs Assy.-Sec.
54. Fuel Bowl-Pri.
55. Fuel Bowl-Sec.
56. Pump Diaphragm Assy.
57. Pump Lever Stud
58. Diaphragm Return Spring
59. Pump Operating Adj. Screw Spring
60. Float Spring
61. Choke Spring
62. Throttle Stop Screw Spring
63. Throttle Lever Ball Nut
64. Fuel Valve Seat Adj. Nut
65. Pump Operating Adj. Nut
66. Pump Cam
67. Pump Operating Lever Screw Sleeve
68. Pivot Screw L.W.
69. Pivot Screw Washer
70. Throttle Lever Ball L.W.
71. Pivot Screw Cotter Pin
72. Connecting Link Cotter Pin
73. Float Shaft Retaining Bracket
74. Fuel Bowl Vent Baffle
75. Pump Operating Lever & Guide Assy.
76. Pump Operating Lever Retainer

Super Tuning and Modifying Holley Carburetors

EXPLODED VIEW - MODEL 4360

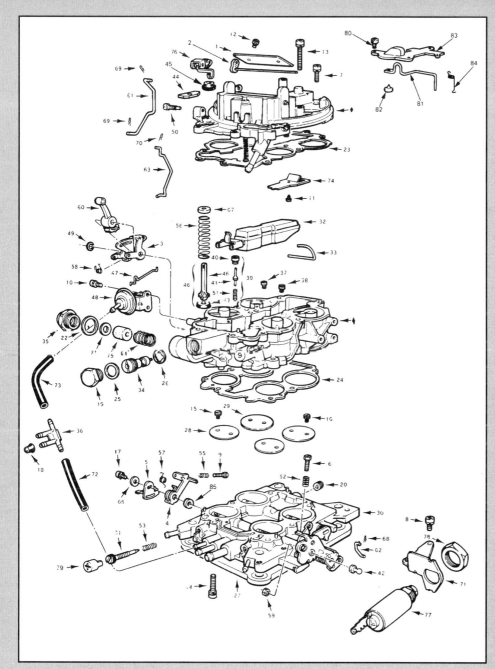

1. Choke Plate
2. Choke Shaft & Lever Assembly
3. Choke Control Lever
4. Fast Idle Cam Lever
5. Dechoke Lever
6. Kill Idle Adjusting Screw
7. Air Horn to Main Body Screw - Short
8. Solenoid Bracket Screw & L.W.
9. Fast Idle Adjusting Screw
10. Choke Diaphragm Bracket Screw
11. Fuel Bowl Baffle Screw
12. Choke Plate Screw
13. Air Horn to Main Body Screw - Long
14. Throttle Body to Main Body Screw & L.W.
15. Throttle Plate Screw Pri.
16. Throttle Plate Screw Sec.
17. Dechoke Lever Screw & L.W.
18. TEE Plug
19. Fuel Inlet Plug
20. Power Brake Plug
21. Fuel Inlet Filter Gasket
22. Fuel Inlet Fitting Gasket
23. Main Body Gasket
24. Throttle Body Gasket
25. Fuel Inlet Plug Gasket
26. Fuel Valve Seat Gasket
27. Flange Gasket
28. Throttle Plate Pri.
29. Throttle Plate Sec.
30. Throttle Body & Shaft Assy.
31. Idle Adjusting Needle
32. Float & Hinge Assy.
33. Float Hinge Shaft & Retainer
34. Fuel Inlet Valve Assy.
35. Fuel Inlet Fitting
36. TEE Connector
37. Main Jet Primary
38. Main Jet Secondary
39. Power Valve Assy.
40. Power Valve Needle Seat
41. Power Valve Needle
42. Throttle Lever Ball
43. Pump Cup
44. Choke Rod Seal
45. Pump Stem Seal
46. Accelerating Pump Assy.
47. Choke Diaphragm Link
48. Choke Diaphragm Assy.
49. Choke Control Lever Ret.
50. Pump Lever Stud
51. Power Valve Spring
52. Kill Idle Screw Spring
53. Idle Needle Spring
54. Fuel Inlet Filter Spring
55. Fast Idle Screw Spring
56. Drive Spring
57. Fast Idle Cam Lever Return Spring
58. Choke Control Lever Spring
59. Throttle Lever Ball Nut
60. Fast Idle Cam Assy.
61. Choke Rod
62. Secondary Connecting Rod
63. Accelerating Pump Rod
64. Throttle Lever Ball L.W.
65. Connecting Rod Washer
66. Dechoke Lever Retaining W.
67. Spring Perch Washer
68. Connecting Rod Retainer
69. Choke Rod Retainer
70. Pump Rod Retainer
71. Solenoid Bracket
72. Choke Vacuum Hose
73. Choke Vacuum Hose
74. Fuel Bowl Baffle
75. Fuel Inlet Filter
76. Accelerating Pump Lever
77. Solenoid Idle Stop
78. Solenoid Nut
79. Idle Adjusting Needle Limiter
80. E.C.S. Vent Cover Screw
81. E.C.S. Vent Rod
82. E.C.S. Vent Seal
83. E.C.S. Vent Cover
84. E.C.S. Vent Return Spring
86. Fast Idle Cam Lever Spacer

Parts not shown on Illustration
P.C.V. Tube Plug
Throttle Lever Ball L.W.
Trans Kick-Down Stud
Trans Kick-Down Nut

Super Tuning and Modifying Holley Carburetors

EXPLODED VIEW - MODEL 4150/4160

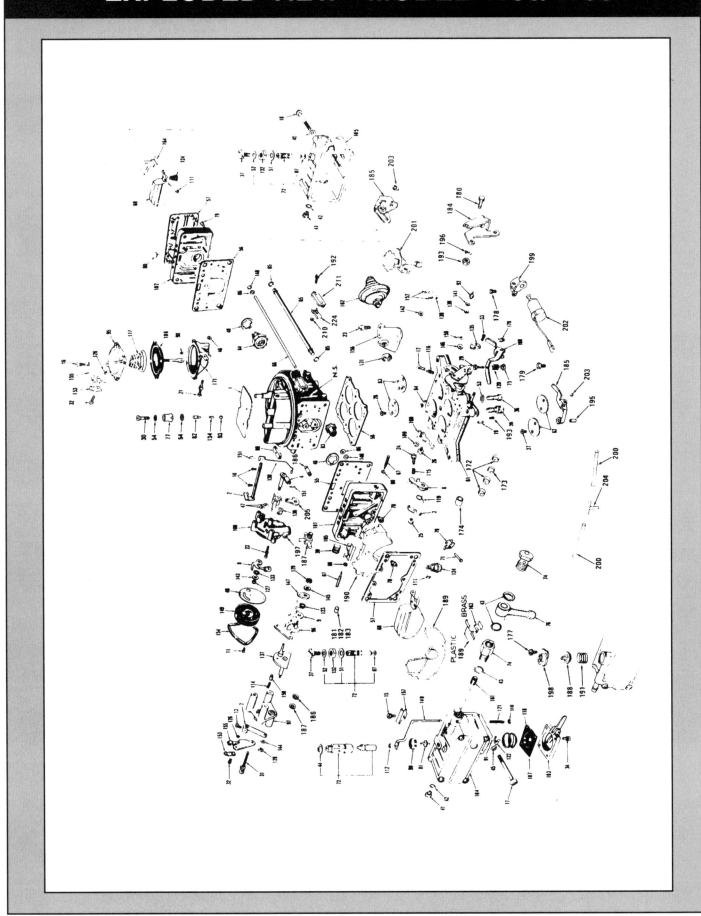

EXPLODED VIEW - MODEL 4150/4160 (CONT.)

#	Part	#	Part	#	Part
1	Choke Plate	70	Float Hinge Adapter	139	Throttle Connecting Rod
2	Choke Shaft Assembly	71	Float Lever Shaft	140	Air Vent Push Rod
3	Fast Idle Pick-up Lever	72	Fuel Inlet Valve & Seat Assy.	141	Throt. Lev. Ball Nut Washer
4	Choke Hsg. Shaft & Lev. Assy.	73	Pump Lever Adj. Scr. Fitting	143	Choke Shaft Nut Lock Washer
5	Choke Control Lever	74	Fuel Inlet Fitting	144	Choke Cont. Lev. Nut L.W.
6	Fast Idle Cam Lever	75	Fuel Transfer Tube Fitting	145	Back-up Plate Stud Nut L.W.
7	Choke Lev. & Swivel Assy.	76	Fuel Inlet Tube & Fitting Assembly	146	Throt. Connector Pin Washer
8	Choke Therm. Lev., Link & Piston Assy.	77	Pump Discharge Nozzle	147	Choke Spring Washer
9	Choke Rod Lev. & Bush Assy.	78	Main Jet-Primary	149	Therm. Housing Assy - Complete
10	Choke Plate Screw	79	Main Jet- Secondary	150	Throttle Connector Pin Retainer
11	Therm. Housing Clamp Screw	80	Bowl Vent Valve	151	Choke Rod Retainer
12	Throttle Stop Screw	81	Air Vent Valve	152	Throt. Connecting Rod Cotter Pin
13	Choke Lev. Assy. Swivel Screw	82	Pump Discharge Needle Valve	153	Choke Cont. Wire Bracket Clamp
14	Choke Diaph. Assy., Brkt. Scr. & L.W.	83	Power Valve Assy - Primary	154	Thermostat Housing Clamp
15	Air Vent Clamp Screw	84	Power Valve Assy - Secondary	155	Choke Control Wire Bracket
16	Sec. Diaph. Assy. Scr. & L.W.	85	Fuel Line Tube "O" Ring Seal	**	Choke Pump Rod Spring
17	Fuel Bowl to Main Body Screw - Primary	87	Fuel Valve Seat "O" Ring Seal	156	Dashpot Bracket
18	Fuel Bowl to Main Body Screw - Secondary	88	Idle Needle Seal	157	Air Vent Rod Clamp
19	Diaph. Lev. Adjusting Screw	89	Choke Rod Seal	158	Fast Idle Cam Plunger
20	Throt. Body Screw & Lock Washer	90	Diaphragm Hsg. Check Ball - Sec.	159	Choke Vacuum Tube
21	Diaph. Hsg. Assy. Screw & Lock	91	Pump Inlet Check Ball or Valve	160	Fuel Transfer Tube
22	Choke Hsg. Screw & L.W.	92	Throt. Lev. Ball	161	Filter Screen
23	Dashpot Brkt. & L.W.	93	Pump Discharge Check Ball	162	Dashpot Assembly
24	Fast Idle Cam Lever Adj. Screw	94	Choke Diaph. Assy. Link	163	Baffle Plate - Primary
25	Fast Idle Cam Lever Screw & L.W.	95	Sec. Diaph. Housing Cover	164	Baffle Plate - Secondary
26	Diaph. Lev. Assy. Scr. & L.W.	96	Back-up Plate & Stud Assy.	165	Metering Body Vent Baffle
27	Throt. Plate Screw - Primary	97	Fast Idle Cam Plate	166	Float Shaft Retainer Bracket
28	Throt. Plate Screw - Secondary	99	Air Vent Cap	167	Fuel Inlet Filter
29	Pump Lever Adj. Screw	100	Choke Hsg. & Plugs Assembly	168	Diaphragm Lever Assembly
30	Pump Discharge Nozzle Screw	101	Main Metering Body & Plugs Assy. - Primary	169	Pump Operating Lever
31	Fast Idle Cam Plate Screw & L.W.	102	Main Metering Body & Plugs Assy. - Secondary	170	Pump Operating Lever Retainer
32	Choke Cont. Wire Brkt. Clamp Screw	103	Fuel Pump Cover Assembly	171	Secondary Diaphragm Housing
33	Pump Cam Lock Screw	104	Fuel Bowl & Plugs Assy. - Primary	172	Throt. Shaft Brg. Pri. & Sec. (Ribbon)
34	Fuel Pump Cov. Assy. Screw & L.W.	105	Fuel Bowl & Plugs Assy. - Secondary	173	Throt. Shaft Brg. Pri. & Sec. (Ribbon)
36	Throt. Body Screw - Special	106	Sec. Diaph. & Rod Assy.	174	Throt. Shaft Brg. Pri. (Solid)
37	Fuel Valve Seat Lock Screw	107	Pump Diaph. Assembly	175	Plug-Fuel Bowl Drain
38	Float Shaft Brkt. Scr. & L.W.	108	Choke Diaph. Assy - Complete	**	Choke Nut Tube
39	Spark Hole Plug	109	Sec. Diaph. Link Retainer	176	Pump Oper. Lever Stud
40	Fuel Bowl Plug	110	Air Vent Rod Spring Retainer	177	Vent Valve Screw & L.W.
41	Fuel Level Check Plug	111	Float Retainer	178	Solenoid Bracket Screw & L.W.
42	Fuel Level Check Plug Gasket	112	Air Vent Valve Retainer	179	Solenoid Bracket Screw & L.W.
43	Fuel Inlet Fitting Gasket	113	Choke Control Lever Retainer	180	Throttle Lever Extension Screw
44	Fuel Valve Seat Gasket	114	Fast Idle Cam Plunger Spring	181	Vacuum Tube Plug
45	Fuel Bowl Screw Gasket	115	Fast Idle Cam Lever Screw Spring	182	Spark Tube Plug
46	Sec. Diaph. Housing Gasket	116	Throt. Stop Screw Spring	183	Emission Tube Plug
47	Choke Housing Gasket	117	Secondary Diaphragm Spring	184	Throttle Lever Extension
48	Power Valve Body Gasket	118	Pump Diaphragm Return Spring	185	Cam Follower Lever
49	Choke Therm. Housing Gasket	119	Fast Idle Cam Lever Spring	186	Tube & "O" Ring Assy.
51	Fuel Valve Seat Adj. Nut Gasket	120	Pump Lev Adj. Screw Spring	187	Four Way Pipe Connector
52	Fuel Valve Seat Lock Screw Gasket	121	Air Vent Rod Spring	188	Vent Valve Body
53	Throt. Body Screw Gasket	122	Pump Inlet Check Ball Ret. Spring	189	Fuel Bowl Filter
54	Pump Discharge Nozzle Gasket	123	Choke Spring	190	Metering Body Filter
55	Metering Body Gasket - Primary	124	Float Spring - Primary & Secondary	191	Vent Valve Spring
56	Metering Body Gasket - Secondary	125	Fuel Inlet Filter Spring Screw Nut	193	Throttle Lever Extension Nut
57	Fuel Bowl Gasket	126	Choke Cont. Wire Brkt. Clamp	194	Solenoid Bracket Screw Nut
58	Throttle Body Gasket	127	Choke Thermostat Shaft Nut	195	Pump Operating Lever Screw Spring
59	Fuel Bowl Plug Gasket	128	Choke Control Lever Nut	196	Throttle Lever Extension L/W.
60	Fuel Inlet Filter Gasket	129	Back-up Plate Stud Nut	197	Dashpot Lock Washer
61	Flange Gasket	130	Throt. Lev Ball Nut	198	Vent Valve Clamp Assembly
62	Throttle Plate - Primary	131	Dashpot Nut	199	Solenoid Bracket
63	Throttle Plate - Secondary	132	Fuel Valve Seat Adjusting Nut	200	Choke Vacuum Hose
64	Throt. Body & Shaft Assy.	133	Choke Thermostat Lever Spacer	201	Modulator Assy.
65	Fuel Line Tube	134	Pump Check Ball Weight	202	Solenoid Assy.
67	Idle Adjusting Needle	135	Pump Cam	203	Cam Follower Assy.
68	Float & Hinge Assy - Primary	136	Fast Idle Cam Assembly	204	"T" Connector
69	Float & Hinge Assy - Secondary	137	Fast Idle Cam & Shaft Assembly	205	Choke Thermostat Lever
		138	Choke Rod	**	Not Shown on Exploded View

Super Tuning and Modifying Holley Carburetors

EXPLODED VIEW - MODEL 2300

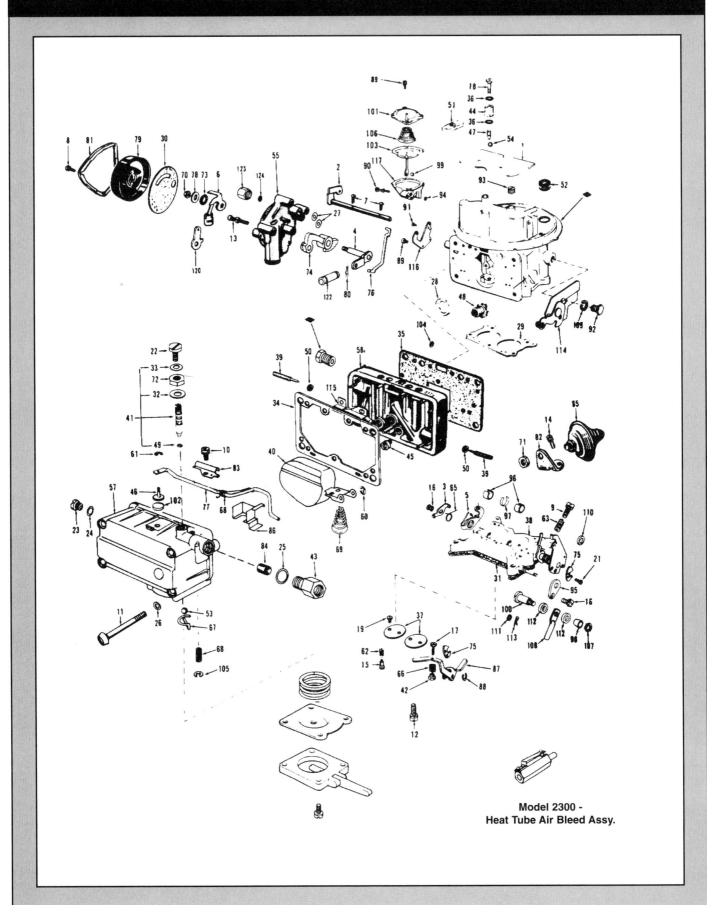

Model 2300 - Heat Tube Air Bleed Assy.

124 *Super Tuning and Modifying Holley Carburetors*

EXPLODED VIEW - MODEL 2300 (CONT.)

1. Choke Plate
2. Choke Shaft Assembly
3. Fast Idle Pick-up Lever
4. Choke Hsg. Shaft & Lev. Assy.
5. Fast Idle Cam Lover
6. Choke Therm. Lev., Link & Piston Assembly
7. Choke Plate Screw
8. Therm. Hsg. Clamp Screw
9. Throttle Stop Screw
10. Air Vent Rod Clamp Scr. & L.W.
11. Fuel Bowl to Main Body Screw
12. Throt. Body Scr. & L.W.
13. Choke Hsg. Scr. & L.W.
14. Dashpot Brkt. Scr. & L.W.
15. Fast Idle Cam Lover Screw
16. Fast Idle Cam Lev. & Throt. Lev. Screw & L.W.
17. Pump Oper. Lev. Adj. Screw
18. Pump Discharge Nozzle Screw
19. Throttle Plate Screw
20. Fuel Pump Cov. Assy. Scr. & L.W.
21. Pump Cam Lock Scr. & L.W.
22. Fuel Valvo Scat Lock Screw
23. Fuel Level Check Plug
24. Fuel Level Check Plug Gasket
26. Fuel Inlet Fitting Gasket
26. Fuel Bowl Screw Gasket
27. Choke Housing Gasket
28. Power Valve Body Gasket
29. Throttle Body Gasket
30. Choke Therm. Housing Gasket
31. Flange Gasket
32. Fuel Valve Seat Adj. Nut Gasket
33. Fuel Valve Seat Lock Scr. Gskt.
34. Fuel Bowl Gasket
35. Metering Body Gasket
36. Pump Discharge Nozzle Gasket
37. Throttle Plate
38. Throt. Body & Shaft Assy.
39. Idle Adjusting Needle
40. Float & Hinge Assy.
41. Fuel Inlet Valve & Seat Assy.
42. Pump Oper. Lev. Adj. Scr. Fitting
43. Fuel Inlet Fitting
44. Pump Discharge Nozzle
45. Main Jet
46. Air Vent Valve
47. Pump Discharge Needle Valve or Check Ball Weight
48. Power Valve Assembly
49. Fuel Valve Seat "O" Ring Seal or Gasket
50. Idle Needle Seal
51. Choke Rod Seal
52. Choke Code Air Tube Grommet
53. Pump Inlet Check Ball or Valve
54. Pump Discharge Check Ball
55. Choke Hsg. & Plugs Assy.
56. Fuel Pump Cover Assy.
67. Fuel Bowl & Plugs Assy.
58. Main Metering Body & Plugs Assy.
59. Pump Diaphragm Assembly
60. Float Spring Retainer
61. Air Vent Retainer
62. Fast Idle Cam Lev. Scr. Spring
63. Throttle Stop Screw Spring
64. Pump Diaphragm Return Spring
65. Fast Idle Cam Lev. Spring
66. Pump Oper. Lev. Adj. Spring
67. Pump Inlet Check Ball Retainer
68. Air Vent Rod Spring
69. Float Spring
70. Choke Thermostat Shaft Nut
71. Dashpot Screw Nut
72. Fuel Valve Seat Adj. Nut
73. Choke Therm. Lover Spacer
74. Fast Idle Cam Assembly
75. Pump Cam
76. Choke Rod
77. Air Vent Rod
78. Choke Therm. Shaft Nut L.W.
79. Thermostat Housing Assembly
80. Choke Rod Retainer
81. Thermostat Housing Clamp
82. Dashpot Bracket
83. Air Vent Rod Clamp
84. Filter Screen Assembly
85. Dashpot Assembly
86. Baffle Plate
87. Pump Operating Lever
88. Pump Operating Lev. Retainer
89. Adapter Mounting & Diaphragm Cover Assy. Screw
90. Throt. Diaphragm Hsg. Scr.
91. Adapter Passage Screw
92. Choke Bracket Screw
93. Air Adapter Hole Plug
94. Throt. Diaphragm Hsg. Gskt.
95. Throttle Lover
96. Throttle Shaft Bearing
97. Throttle Shaft Brg. (Center)
98. Throttle Connector Pin Bushing
99. Diaphragm Check Ball
100. Throttle Connector Pin
101. Diaphragm Housing Cover
102. Air Vent Cap
103. Diaphragm Assembly
104. Diaphragm Link Retainer
105. Air Vent Rod Spg. Retainer
106. Diaphragm Spring
107. Throttle Link Connector Pin Nut
108. Throttle Connector Bar
109. Choke, Brkt. Scr. Lock Washer
110. Throt. Link Connector Pin
111. Throt. Connector Pin Washer
112. Throt. Connector Pin Spacer
113. Throt. Connector Pin Retainer
114. Choke Control Lever Bracket
115. Metering Body Vent Baffle
116. Throt. Diaphragm Adapter
117. Diaphragm Housing
118. Idle Adj. Needle Spring
119. Pump Oper. Lever Stud
   * Vent Tubs
   * Heat Tube Nut
   * Heat Tube Ferrule
   * Fuel Tube Hose Clamp
   * Retainer
   * Air Cleaner Stud (Long)
   * Air Cleaner Stud (Short)
   * Choke Heat Tube
   * Fuel Line Hose
   * Fresh Air Hose
   * Spring
120. Choke Thermostat Lever
121. Vacuum Tube Plug
122. Tube & "O" Ring Assy.
123. Nut
124. Screen

Parts not shown
   Solenoid Bracket
   Screw & L.W.
   Bracket Screw
   Bracket Nut
   Solenoid Bracket Assy.
* Not Shown on Exploded View

Super Tuning and Modifying Holley Carburetors

NUMERICAL LISTING AND COMPONENT PARTS

CARBURETOR PART NO.	CARB. MODEL NO.	CFM	RENEW KIT	TRICK KIT	PRIMARY & SECONDARY NEEDLE & SEAT	PRIMARY MAIN JET	SECONDARY MAIN JET OR PLATE	PRIMARY METERING BLOCK	SECONDARY METERING BLOCK	PRIMARY POWER VALVE	PRIMARY DISCHARGE NOZZLE SIZE
R1848-1	4160	465	37-119	37-912	6-506	122-57	134-31	134-125	134-3	125-85	.025
R1849	4160	550	37-119	37-912	6-506	122-62	134-5	134-126	134-5	125-85	.025
R1850-2	4160	600	37-119	37-912	6-506	122-66	134-9	134-128	134-9	125-65	.025
R1850-3	4160	600	37-119	37-912	6-506	122-66	134-9	134-128	134-9	125-65	.025
R1850-4	4160	600	37-119	37-912	6-506	122-66	134-9	134-128	134-9	125-65	.031
R2818-1	4150	600	37-333	37-905	6-506	122-65	122-76	134-129	134-130	125-65	.025
R3124	4150	750	37-1184	N/A	6-504	122-70	122-76	N/S	N/S	125-85(12)	.025
R3247	4150	780	37-1184	N/A	6-504	122-70	122-76	N/S	134-132	125-85(12)	.021
R3259-1	4150	725	1085-2489	N/A	6-73	122-68	122-78	N/S	N/S	125-85	.025
R3310-1	4150	780	37-1084	37-915	6-504	122-72	122-76	134-131	134-133	(12,13)	.025
R3310-2	4160	750	37-754	37-921	6-504	122-72	134-21	134-131	134-21	125-65	.025
R3310-3	4160	750	37-754	37-921	6-504	122-72	134-21	134-131	134-21	125-65	.025
R3310-4	4160	750	37-754	37-921	6-504	122-72	134-21	134-131	134-21	125-65	.031
R3367	4160	585	37-119	N/A	6-507	122-65	134-22	N/S	N/R	125-65	.025
R3370	4160	585	37-119	N/A	6-504	122-65	134-17	N/S	N/R	125-65	.025
R3418-1	4150	855	37-1273	N/A	6-504	78C/82T	82C/80T	N/S	N/S	(15,21)	.028
R3613	4150	770	37-1184	N/A	6-504	122-71	122-76	N/S	134-132	125-85(12)	.021
R3659	2300	466	37-396	N/A	6-504	N/R	N/S	N/R	N/S	N/R	N/R
R3660	2300	350	37-396	N/A	6-504	122-64	N/R	N/S	N/R	125-65	.021
R3807	4150	595	37-395	N/A	6-516	122-67	122-72	N/S	N/S	125-65	.025
R3810	4160	585	37-395	N/A	6-516	122-65	134-72	N/S	N/S	125-65	.025
R3811	4160	585	37-395	N/A	6-516	122-65	134-17	N/S	N/R	125-65	.025
R3910	4150	780	37-1184	N/A	6-504	122-71	122-76	N/S	N/S	125-65(12)	.021
R4053	4150	780	37-1184	N/A	6-504	122-68	122-76	N/S	N/S	125-65(12)	.025
R4055-1	2300	350	37-396	N/A	6-504	122-63	N/R	N/S	N/R	125-65	.021
R4056-1	2300	350	37-396	N/A	6-504	122-61	N/R	N/S	N/R	125-65	.025
R4118	4150	725	37-1184	37-906	6-504	122-68	122-78	134-134	134-135	125-85	.025
R4144-1	2300	350	37-396	N/A	6-504	122-62	N/R	1034-6020	N/R	125-65	.031
R4224	4160	660	37-424	N/A	6-508	122-76	134-12	134-136	134-12	N/R	.025
R4235	4160	770	37-485	N/A	6-504	(29)	N/S	N/S	N/R	125-65	.035
R4236	4160	770	37-485	N/A	6-504	122-80	N/S	N/S	N/R	125-65	.035
R4295	4150	585	37-485	N/A	6-504	122-69	122-71	N/S	N/S	125-65	.025
R4296	4150	850	37-485	N/A	6-504	78C/82T	82C/80T	N/S	N/S	125-65(15)	.035
R4346	4150	780	37-1184	N/A	6-504	122-68	122-76	N/S	N/S	125-85(12)	.025
R4365-1	2300	500	37-396	N/A	6-504	N/R	1034-6153-35	N/R	N/R	N/R	N/R
R4412	2300	500	37-474	37-901	6-504	122-73	N/R	134-137	N/R	125-50	.028
R4412-1	2300	500	37-474	37-901	6-504	122-73	N/R	134-137	N/R	125-50	.028
R4412-2	2300	500	37-474	37-901	6-504	122-73	N/R	134-137	N/R	125-50	.028
R4452-1	4160	600	37-119	37-912	6-506	122-63	134-39	134-138	134-39	125-85	.031
R4473	4160	450	703-1	N/A	6-506	122-58	N/S	N/S	N/S	125-85	.025
R4490	4150	780	37-1184	N/A	6-504	122-70	122-76	N/S	N/S	125-85(12)	.025
R4514-1	4150	700	37-422	N/A	6-504	122-66	122-79	N/S	N/S	125-65	.029
R4548	4160	450	37-119	N/A	6-506	122-57	134-30	134-139	134-130	1025-219-65	.031
R4555	4150	780	37-1184	N/A	6-504	122-70	122-76	N/S	N/S	125-85(12)	.025
R4575	4500	1050	37-487	37-920	6-504	122-84	122-84	134-140	134-141	125-65(15)	.035
R4609	4150	730	37-422	N/A	6-504	122-66	122-79	N/S	N/S	125-65	.029
R4628	4150	780	37-422	N/A	6-504	122-70	122-83	N/S	N/S	125-85	.026
R4647	4150	735	37-422	N/A	6-504	122-64	122-82	N/S	N/S	125-85	.031
R4653	4150	780	37-422	N/A	6-504	122-71	122-82	N/S	N/S	125-65	.026
R4670	2300	350	37-396	N/A	6-504	122-62	N/R	1034-6020	N/R	125-65	.031
R4672	2300	500	37-396	N/A	6-504	N/R	1034-6153-35	N/R	N/R	N/R	N/R
R4691-2	2110	300	37-496	N/A	6-509	122-63	N/R	N/R	N/R	N/R	.021
R4742	4150	600	37-1184	N/A	6-504	122-63	122-72	N/S	N/S	N/S	.031
R4776	4150	600	37-485	37-910	6-504	122-69	122-71	134-142	134-143	125-65	.025
R4776-1	4150	600	37-485	37-910	6-504	122-66	122-76	134-144	134-143	125-65	.028
R4776-2	4150	600	37-485	37-910	6-504	122-66	122-76	134-145	134-143	125-65	.028
R4776-3	4150	600	37-485	37-910	6-504	122-66	122-73	134-145	134-143	125-65	.028
R4776-4	4150	600	37-485	37-910	6-504	122-66	122-73	134-145	134-143	125-65	.028
R4777	4150	650	37-485	37-910	6-504	122-71	122-76	134-146	134-147	125-65	.025
R4777-1	4150	650	37-485	37-910	6-504	122-67	122-76	134-148	134-147	125-65	.028
R4777-2	4150	650	37-485	37-910	6-504	122-67	122-76	134-150	134-147	125-65	.028
R4777-3	4150	650	37-485	37-910	6-504	122-67	122-73	134-150	134-147	125-65	.028
R4777-4	4150	650	37-485	37-910	6-504	122-67	122-73	134-150	134-147	125-65	.028

NUMERICAL LISTING AND COMPONENT PARTS

SECONDARY NOZZLE SIZE OR SPRING COLOR	PRIMARY BOWL GASKET†	PRIMARY METERING BLOCK GASKET†	SECONDARY BOWL GASKET†	SECONDARY METERING BLOCK GASKET†	SECONDARY METERING PLATE GASKET†	PRIMARY FUEL BOWL	SECONDARY FUEL BOWL	THROTTLE BODY & SHAFT ASSEMBLY	VENTURI DIAMETER PRIMARY	VENTURI DIAMETER SECONDARY	THROTTLE BORE DIAMETER PRIMARY	THROTTLE BORE DIAMETER SECONDARY
Green	108-33	108-29	108-30	108-30	108-27	134-100	134-105	112-37	1-3/32	1-3/32	1-1/2	1-1/2
Plain	108-33	108-29	108-30	108-30	108-27	134-100	134-105	112-38	1-3/16	1-1/4	1-1/2	1-1/2
Plain	108-33	108-29	108-30	108-30	108-27	134-101	134-105	112-20	1-1/4	1-5/16	1-9/16	1-9/16
Plain	108-33	108-29	108-30	108-30	N/R	134-101	134-105	112-20	1-1/4	1-5/16	1-9/16	1-9/16
Plain	108-33	108-29	108-30	108-30	N/R	134-101	134-105	112-20	1-1/4	1-5/16	1-9/16	1-9/16
Purple	108-33	108-29	108-33	108-29	N/R	134-101	134-106	112-39	1-1/4	1-5/16	1-9/16	1-9/16
Yellow	108-33	108-29	108-33	108-29	N/R	N/S	N/S	N/S	1-3/8	1-7/16	1-11/16	1-11/16
Yellow	108-33	108-29	108-33	108-29	N/R	N/S	134-102	N/S	1-3/8	1-7/16	1-11/16	1-11/16
Yellow	108-33	108-29	108-33	108-29	N/R	34-14	34-14	N/S	1-5/16	1-3/8	1-11/16	1-11/16
Plain	108-33	108-29	108-33	108-29	N/R	134-103	134-102	112-9	1-3/8	1-7/16	1-11/16	1-11/16
Plain	108-33	108-29	108-30	108-30	108-27	134-103	134-102	112-9	1-3/8	1-7/16	1-11/16	1-11/16
Plain	108-33	108-29	108-30	108-30	108-27	134-103	134-102	112-9	1-3/8	1-7/16	1-11/16	1-11/16
Plain	108-33	108-29	108-30	108-30	108-27	134-103	134-102	112-9	1-3/8	1-7/16	1-11/16	1-11/16
Purple	108-33	108-29	108-30	108-30	108-27	N/S	134-105	N/S	1-1/4	1-5/16	1-9/16	1-9/16
Purple	108-33	108-29	108-30	108-30	108-27	N/S	134-105	N/S	1-1/4	1-5/16	1-9/16	1-9/16
Yellow	108-33	108-29	108-33	108-29	N/R	134-103	1034-5972	N/S	1-9/16	1-9/16	1-3/4	1-3/4
Yellow	108-33	108-29	108-33	108-29	N/R	1034-4703	1034-4652	N/S	1-3/8	1-7/16	1-11/16	1-11/16
Brown	N/R	N/R	108-30	108-30	108-27	N/R	1034-5643	N/S	1-3/8	N/A	1-3/4	N/A
N/R	100-33	108-29	N/R	N/R	N/H	1034-5058	N/R	N/S	1-3/16	N/A	1-1/2	N/A
Purple	108-33	108-29	108-33	108-29	N/R	N/S	1034-5972	N/S	1-1/4	1-5/16	1-9/16	1-9/16
Purple	108-33	108-29	108-30	108-30	108-27	N/S	1034-5972	N/S	1-1/4	1-5/16	1-9/16	1-9/16
Purple	108-33	108-29	108-30	108-30	108-27	N/S	1034-5972	N/S	1-1/4	1-5/16	1-9/16	1-9/16
Yellow	108-33	108-29	108-33	108-29	N/R	134-103	1034-4652	N/S	1-3/8	1-7/16	1-11/16	1-11/16
Yellow	108-33	108-29	108-33	108-29	N/R	134-103	1034-4652	N/S	1-3/8	1-7/16	1-11/16	1-11/16
N/R	108-33	108-29	N/R	N/R	N/S	N/S	N/R	N/S	1-3/16	N/A	1-1/2	N/A
N/R	108-33	108-29	N/R	N/R	N/S	N/S	N/R	N/S	1-3/16	N/A	1-1/2	N/A
Yellow	108-33	108-29	108-33	108-29	N/R	134-103	134-102	112-41	1-5/16	1-3/8	1-11/16	1-11/16
N/R	108-33	108-29	N/R	N/R	N/S	N/S	N/R	N/S	1-3/16	N/A	1-1/2	N/A
.025	108-33	108-29	108-30	108-30	108-27	134-101	134-106	112-3	1-1/4	1-5/16	1-11/16	1-11/16
Plain	108-33	108-29	108-30	108-30	108-27	134-101	134-106	N/S	1-3/8	1-7/16	1-11/16	1-11/16
Plain	108-33	108-29	108-30	108-30	108-27	134-101	134-106	N/S	1-3/8	1-7/16	1-11/16	1-11/16
.025	108-33	108-29	108-33	108-29	N/R	N/S	N/S	N/S	1-1/4	1-5/16	1-9/16	1-9/16
.035	108-33	108-29	108-33	108-29	N/R	134-103	N/S	N/S	1-9/16	1-9/16	1-3/4	1-3/4
Yellow	108-33	108-29	108-33	108-29	N/R	N/S	1034-4652	N/S	1-3/8	1-7/16	1-11/16	1-11/16
Yellow	N/R	N/R	108-30	108-30	108-13	1034-5892	N/R	N/S	1-9/16	N/A	1-3/4	N/A
N/R	108-33	108-29	N/R	N/R	N/R	134-103	N/R	112-2	1-3/8	N/R	1-11/16	N/R
N/R	108-33	108-29	N/R	N/R	N/R	134-103	N/R	112-2	1-3/8	N/R	1-11/16	N/R
N/R	108-33	108-29	N/R	N/R	N/R	134-103	N/R	112-2	1-3/8	N/R	1-11/16	N/R
Purple	108-33	108-29	108-30	108-30	108-27	134-100	134-105	112-42	1-1/4	1-5/16	1-9/16	1-9/16
Orange	108-33	108-29	108-30	108-30	108-27	N/S	N/S	N/S	1-3/32	1-3/32	1-1/2	1-1/2
Yellow	108-33	108-29	108-33	108-29	N/R	134-103	1034-4652	N/S	1-3/8	1-7/16	1-11/16	1-11/16
Yellow	108-33	108-29	108-33	108-29	N/R	N/S	1034-4652	N/S	1-3/8	1-7/16	1-11/16	1-11/16
Brown	108-33	108-29	108-30	108-30	108-27	134-100	134-105	112-43	1-3/32	1-3/32	1-1/2	1-1/2
Yellow	108-33	108-29	108-33	108-29	N/R	134-103	1034-4652	N/S	1-3/8	1-7/16	1-11/16	1-11/16
.035	108-33	108-29	108-33	108-29	N/R	134-108	134-112	N/S	1-11/16	1-11/16	2	2
Yellow	108-33	108-29	108-33	108-29	N/R	1034-5694	1034-4652	N/S	1-3/8	1-7/16	1-11/16	1-11/16
Yellow	108-33	108-29	108-33	108-29	N/R	N/S	1034-4652	N/S	1-3/8	1-7/16	1-11/16	1-11/16
Purple	108-33	108-29	108-33	108-29	N/R	N/S	1034-4652	N/S	1-3/8	1-7/16	1-11/16	1-11/16
Purple	108-33	108-29	108-33	108-29	N/R	N/S	1034-4652	N/S	1-3/8	1-7/16	1-11/16	1-11/16
N/R	108-33	108-29	N/R	N/R	N/R	N/S	N/R	N/S	1-3/16	N/A	1-1/2	N/A
Yellow	N/R	N/R	108-30	108-30	108-13	N/R	1034-5892	N/S	1-9/16	N/A	1-3/4	N/A
N/R	N/R	N/R	N/R	N/R	N/R	N/R	N/R	N/R	1-5/32	N/R	1-7/16	N/R
Purple	108-33	108-29	108-33	108-29	N/R	N/S	N/S	N/S	1-1/4	1-5/16	1-9/16	1-9/16
.032	108-33	108-29	108-33	108-29	N/R	134-103	134-104	112-6	1-1/4	1-5/16	1-9/16	1-9/16
.032	108-33	108-29	108-33	108-29	N/R	134-103	134-104	112-24	1-1/4	1-5/16	1-9/16	1-9/16
.032	108-33	108-29	108-33	108-29	N/R	134-103	134-104	112-16	1-1/4	1-5/16	1-9/16	1-9/16
.032	108-33	108-29	108-33	108-29	N/R	134-103	134-104	112-16	1-1/4	1-5/16	1-9/16	1-9/16
.032	108-33	108-29	108-33	108-29	N/R	134-103	134-104	112-16	1-1/4	1-5/16	1-9/16	1-9/16
.025	108-33	108-29	108-33	108-29	N/R	134-103	134-104	112-7	1-1/4	1-5/16	1-11/16	1-11/16
.028	108-33	108-29	108-33	108-29	N/R	134-103	134-104	112-13	1-1/4	1-5/16	1-11/16	1-11/16
.028	108-33	108-29	108-33	108-29	N/R	134-103	134-104	112-17	1-1/4	1-5/16	1-11/16	1-11/16
.028	108-33	108-29	108-33	108-29	N/R	134-103	134-104	112-17	1-1/4	1-5/16	1-11/16	1-11/16
.028	108-33	108-29	108-33	108-29	N/R	134-103	134-104	112-17	1-1/4	1-5/16	1-11/16	1-11/16

NUMERICAL LISTING AND COMPONENT PARTS

CARBURETOR PART NO.	CARB. MODEL NO.	CFM	RENEW KIT	TRICK KIT	PRIMARY & SECONDARY NEEDLE & SEAT	PRIMARY MAIN JET	SECONDARY MAIN JET OR PLATE	PRIMARY METERING BLOCK	SECONDARY METERING BLOCK	PRIMARY POWER VALVE	PRIMARY DISCHARGE NOZZLE SIZE
R4778	4150	700	37-485	37-910	6-504	122-66	122-71	134-151	134-149	125-65	.025
R4778-1	4150	700	37-485	37-910	6-504	122-66	122-76	N/S	134-149	125-65	.028
R4778-2	4150	700	37-485	37-910	6-504	122-66	122-76	134-152	134-149	125-65	.028
R4778-3	4150	700	37-485	37-910	6-504	122-69	122-78	134-225	134-226	125-65	.028
R4778-4	4150	700	37-485	37-910	6-504	122-69	122-78	134-225	134-226	125-65	.028
R4779	4150	750	37-485	37-910	6-504	122-75	122-76	134-153	134-154	125-85	.025
R4779-1	4150	750	37-485	37-910	6-504	122-70	122-80	N/S	134-154	125-85	.028
R4779-2	4150	750	37-485	37-910	6-504	122-70	122-80	134-155	134-154	125-65	.028
R4779-3	4150	750	37-485	37-910	6-504	122-70	122-73	134-227	134-228	125-65	.028
R4779-4	4150	750	37-485	37-910	6-504	122-70	122-80	134-227	134-228	125-65	.028
R4779-5	4150	750	37-485	37-910	6-504	122-70	122-80	134-227	134-228	125-65	.028
R4779-6	4150	750	37-485	37-910	6-504	122-71	122-80	134-227	134-250	125-65	.028
R4780	4150	800	37-485	37-910	6-504	122-72	122-76	134-156	134-157	(12,21)	.031
R4780-1	4150	800	37-485	37-910	6-504	122-70	122-76	N/S	N/S	(12,21)	.031
R4780-2	4150	800	37-485	37-910	6-504	122-70	122-85	134-158	134-159	125-65	.031
R4780-3	4150	800	37-485	37-910	6-504	122-71	122-85	134-229	134-230	125-65	.031
R4780-4	4150	800	37-485	37-910	6-504	122-71	122-85	134-229	134-230	125-65	.031
R4781	4150	850	37-485	37-916	6-504	122-80	122-80	134-160	134-161	125-65(15)	.035
R4781-1	4150	850	37-485	37-916	6-504	122-80	122-80	134-162	134-161	125-65(15)	.031
R4781-2	4150	850	37-485	37-916	6-504	122-80	122-80	134-163	134-161	125-65(15)	.031
R4781-3	4150	850	37-485	37-916	6-504	122-80	122-78	134-231	134-161	125-65(15)	.031
R4781-4	4150	850	37-485	37-916	6-504	122-80	122-78	134-231	134-161	125-65(15)	.031
R4781-5	4150	850	37-485	37-916	6-504	122-80	122-78	134-249	134-212	125-65(15)	.031
R4782	2300	355	37-396	N/A	6-504	122-64	N/R	N/S	N/R	125-65	.031
R4783	2300	500	37-396	N/A	6-504	122-82	N/R	N/R	N/S	N/R	.028
R4788	4150	830	37-485	37-916	6-504	122-80	122-80	134-164	134-165	125-65(B)	.031
R4788-1	4150	830	37-485	37-916	6-504	122-80	122-80	134-164	134-165	125-65(B)	.031
R4790	2300	500	37-396	N/A	6-504	N/R	1034-6153-41	N/S	N/R	N/R	N/R
R4791	2300	350	37-396	N/A	6-504	122-62	N/R	N/S	N/R	125-65	.031
R4792	2300	350	37-396	N/A	6-504	122-61	N/R	N/S	N/R	125-65	.031
R4800-1	4150	780	37-1184	N/A	6-504	122-70	122-76	N/S	N/S	125-85(12)	.025
R4801-1	4150	780	37-1184	N/A	6-504	122-70	122-76	N/S	N/S	125-85(12)	.025
R4802-1	4150	780	37-1184	N/A	6-504	122-70	122-76	N/S	N/S	125-85(12)	.025
R4803-1	4150	780	37-1184	N/A	6-504	122-70	122-76	N/S	N/S	125-85(12)	.025
R6105	2300	500	37-396	N/A	6-504	N/R	N/S	N/S	N/S	N/R	N/R
R6105-1	2300	500	37-396	N/A	6-504	N/R	N/S	N/S	N/S	N/R	N/R
R6106	2300	350	37-396	N/A	6-504	122-65	N/R	N/S	N/S	125-65	.031
R6107	2300	500	37-396	N/A	6-504	N/R	N/S	N/R	N/S	N/R	N/R
R6107-1	2300	500	37-396	N/A	6-504	N/R	N/S	N/R	N/S	N/R	N/R
R6109	4150	750	37-485	37-910	6-504	122-75	122-76	134-242	134-243	125-85	.025
R6129	4150	780	3-422	N/A	6-504	122-70	122-82	N/S	N/S	125-65	.026
R6150	2300	300	3-888	N/A	6-506	122-59	N/R	N/S	N/R	125-25	.028
R6151	4160	600	703-1	N/A	6-506	122-66	134-3	N/S	N/S	125-105	.025
R6152	4160	600	703-1	N/A	6-506	122-66	134-36	N/S	N/S	125-85	.025
R6210-1	4165	650	37-605	37-918	(16,17)	122-602	122-632	134-168	134-169	(14,15)	.025
R6210-2	4165	650	37-605	37-918	(16,17)	122-602	122-83	134-168	134-170	125-85	.025
R6210-3	4165	650	37-605	37-918	(16,17)	122-602	122-83	134-168	134-170	125-85	.025
R6211	4165	800	37-605	37-918	(16,17)	122-62	122-85	134-168	134-171	(14,15)	.025
R6211-1	4165	800	37-605	37-918	(16,17)	122-602	122-85	134-168	134-172	(14,15)	.025
R6212	4165	800	37-606	37-919	6-504	122-63	122-86	134-173	134-174	(14,15)	.025
R6213	4165	800	37-606	37-919	6-504	122-62	122-85	134-173	134-174	(14,15)	.025
R6214	4500	1150	37-608	37-914	6-504	122-95	122-95	134-175	134-175	N/R	.026
R6238-1	4150	780	37-1184	N/A	6-504	122-68	122-73	N/S	N/S	125-65(12)	.025
R6239-1	4150	780	37-1184	N/A	6-504	122-68	122-73	N/S	N/S	125-65(12)	.025
R6244-1	2110	200	37-496	N/A	6-509	122-47	N/R	N/R	N/R	N/R	.021
R6262	4165	800	37-605	37-918	(16,17)	122-62	122-85	134-166	134-166	(14,15)	.025
R6270-1	4160	600	37-397	N/A	N/S	122-64	N/S	N/S	N/S	125-85	.032
R6291	4160	600	37-119	N/A	6-506	122-62	134-39	N/S	N/S	125-85	.031
R6299-1	4160	390	37-654	37-911	6-506	122-50	134-34	N/S	N/S	N/A	.025
R6317	2300	300	3-714	N/A	6-506	122-60	N/R	N/S	N/S	125-50	.028
R6317-1	2300	300	3-714	N/A	6-506	122-60	N/R	N/R	N/S	125-50	.028
R6361	4150	650	703-2	N/A	6-504	122-72	122-84	N/S	N/S	125-85	.026
R6407	4160	450	703-1	N/A	6-506	122-58	134-5	N/S	N/R	125-85	.021

NUMERICAL LISTING AND COMPONENT PARTS

SECONDARY NOZZLE SIZE OR SPRING COLOR	PRIMARY BOWL GASKET†	PRIMARY METERING BLOCK GASKET†	SECONDARY BOWL GASKET†	SECONDARY METERING BLOCK GASKET†	SECONDARY METERING PLATE GASKET†	PRIMARY FUEL BOWL	SECONDARY FUEL BOWL	THROTTLE BODY & SHAFT ASSEMBLY	VENTURI DIAMETER PRIMARY	VENTURI DIAMETER SECONDARY	THROTTLE BORE DIAMETER PRIMARY	THROTTLE BORE DIAMETER SECONDARY
.032	108-33	108-29	108-33	108-29	N/R	134-103	134-104	112-7	1-5/16	1-3/8	1-11/16	1-11/16
.031	108-33	108-29	108-33	108-29	N/R	134-103	134-104	112-25	1-5/16	1-3/8	1-11/16	1-11/16
.031	108-33	108-29	108-33	108-29	N/R	134-103	134-104	112-22	1-5/16	1-3/8	1-11/16	1-11/16
.031	108-33	108-29	108-33	108-29	N/R	134-103	134-104	112-22	1-5/16	1-3/8	1-11/16	1-11/16
.032	108-33	108-29	108-33	108-29	N/R	134-103	134-104	112-5	1-3/8	1-3/8	1-11/16	1-11/16
.031	108-33	108-29	108-33	108-29	N/R	134-103	134-104	112-26	1-3/8	1-3/8	1-11/16	1-11/16
.031	108-33	108-29	108-33	108-29	N/R	134-103	134-104	112-18	1-3/8	1-3/8	1-11/16	1-11/16
.031	108-33	108-29	108-33	108-29	N/R	134-103	134-104	112-18	1-3/8	1-3/8	1-11/16	1-11/16
.031	108-33	108-29	108-33	108-29	N/R	134-103	134-104	112-18	1-3/8	1-3/8	1-11/16	1-11/16
.031	108-33	108-29	108-33	108-29	N/R	134-103	134-104	112-96	1-3/8	1-3/8	1-11/16	1-11/16
.031	108-33	108-29	108-33	108-29	N/R	134-103	134-104	112-8	1-3/8	1-7/16	1-11/16	1-11/16
.031	108-33	108-29	108-33	108-29	N/R	134-103	134-104	112-27	1-3/8	1-7/16	1-11/16	1-11/16
.031	108-33	108-29	108-33	108-29	N/R	134-103	134-104	112-21	1-3/8	1-7/16	1-11/16	1-11/16
.031	108-33	108-29	108-33	108-29	N/R	134-103	134-104	112-21	1-3/8	1-7/16	1-11/16	1-11/16
.025	108-33	108-29	108-33	108-29	N/R	134-103	134-104	112-4	1-9/16	1-9/16	1-3/4	1-3/4
.031	108-33	108-29	108-33	108-29	N/R	134-103	134-104	112-28	1-9/16	1-9/16	1-3/4	1-3/4
.031	108-33	108-29	108-33	108-29	N/R	134-103	134-104	112-19	1-9/16	1-9/16	1-3/4	1-3/4
.031	108-33	108-29	108-33	108-29	N/R	134-103	134-104	112-19	1-9/16	1-9/16	1-3/4	1-3/4
.031	108-33	108-29	108-33	108-29	N/R	134-103	134-104	112-19	1-9/16	1-9/16	1-3/4	1-3/4
.031	108-33	108-29	108-33	108-29	N/R	134-103	134-104	112-97	1-9/16	1-9/16	1-3/4	1-3/4
N/R	108-33	108-29	N/R	N/R	N/R	N/S	N/R	N/S	1-3/16	N/R	1-1/2	N/R
N/R	108-33	108-29	N/R	N/R	N/R	N/S	N/R	N/S	1-9/16	N/R	1-3/4	N/R
.031	108-33	108-29	108-33	108-29	N/R	134-103	134-104	112-44	1-9/16	1-9/16	1-11/16	1-11/16
.031	108-33	108-29	108-33	108-29	N/R	134-103	134-104	112-44	1-9/16	1-9/16	1-11/16	1-11/16
Yellow	N/R	N/R	108-30	108-30	108-13	N/R	1034-5892	N/S	1-9/16	N/R	1-3/4	N/R
N/R	108-33	108-29	N/R	N/R	N/R	N/S	N/S	N/S	1-3/16	N/R	1-1/2	N/R
N/R	108-33	108-29	N/R	N/R	N/R	N/S	N/S	N/S	1-3/16	N/R	1-1/2	N/R
Yellow	108-33	108-29	108-33	108-29	N/R	134-103	1034-4652	N/S	1-3/8	1-7/16	1-11/16	1-11/16
Yellow	108-33	108-29	108-33	108-29	N/R	134-103	1034-4652	N/S	1-3/8	1-7/16	1-11/16	1-11/16
Yellow	108-33	108-29	108-33	108-29	N/R	134-103	1034-4652	N/S	1-3/8	1-7/16	1-11/16	1-11/16
1038-825	N/R	N/R	108-30	108-30	108-13	N/R	N/S	N/S	1-9/16	N/R	1-3/4	N/R
1038-825	N/R	N/R	108-30	108-30	108-13	N/R	N/S	N/S	1-9/16	N/R	1-3/4	N/R
N/R	108-33	108-29	N/R	N/R	N/R	N/S	N/S	N/S	1-3/16	N/R	1-1/2	N/R
1038-825	N/R	N/R	108-30	108-30	108-13	N/R	N/S	N/S	1-9/16	N/R	1-3/4	N/R
1038-825	N/R	N/R	108-30	108-30	108-13	N/R	N/S	N/S	1-9/16	N/R	1-3/4	N/R
.032	108-33	108-29	108-33	108-29	N/R	134-244	134-245	112-29	1-3/8	1-3/8	1-11/16	1-11/16
Purple	108-33	108-29	108-33	108-29	N/R	N/S	N/S	N/S	1-3/8	1-7/16	1-11/16	1-11/16
N/R	108-33	108-29	N/R	N/R	N/R	N/S	N/S	N/S	1-3/16	N/R	1-1/2	N/R
Purple	108-33	108-29	108-30	108-30	108-27	N/S	N/S	N/S	1-1/4	1-5/16	1-9/16	1-9/16
Purple	108-33	108-29	108-30	108-30	108-27	N/S	N/S	N/S	1-1/4	1-5/16	1-9/16	1-9/16
.037	108-32	108-31	108-32	108-31	N/R	134-110	134-114	112-46	1-5/32	1-3/8	1-3/8	2
.037	108-32	108-31	108-32	108-31	N/R	134-110	134-114	112-46	1-5/32	1-3/8	1-3/8	2
.037	108-32	108-31	108-32	108-31	N/R	134-110	134-114	112-47	1-5/32	1-3/8	1-3/8	2
.037	108-32	108-31	108-32	108-31	N/R	134-110	134-114	112-45	1-5/32	1-23/32	1-3/8	2
.037	108-32	108-31	108-32	108-31	N/R	134-110	134-114	112-48	1-5/32	1-23/32	1-3/8	2
.037	108-32	108-31	108-32	108-31	N/R	134-111	134-113	112-45	1-5/32	1-23/32	1-3/8	2
.037	108-32	108-31	108-32	108-31	N/R	134-111	134-113	112-45	1-5/32	1-23/32	1-3/8	2
.026	108-33	108-36	108-33	108-36	N/R	134-108	134-112	N/S	1-13/16	1-13/16	2	2
Yellow	108-33	108-29	108-33	108-29	N/R	N/S	N/S	N/S	1-3/8	1-7/16	1-11/16	1-11/16
Yellow	108-33	108-29	108-33	108-29	N/R	N/S	N/S	N/S	1-3/8	1-7/16	1-11/16	1-11/16
N/R	N/R	N/R	N/R	N/R	N/R	N/R	N/R	N/S	1-5/16	N/R	1-7/16	N/R
.037	108-32	108-31	108-32	108-31	N/R	134-110	134-114	112-45	1-13/16	1-23/32	1-3/8	2
Orange	108-33	108-34	108-30	108-30	108-27	N/S	N/S	N/S	1-1/4	1-5/16	1-9/16	1-9/16
Purple	108-33	108-31	108-30	108-30	108-27	N/S	N/S	N/S	1-1/4	1-5/16	1-5/16	1-9/16
Plain	108-33	108-29	108-30	108-27	108-28	N/S	N/S	N/S	1-1/16	1-1/16	1-7/16	1-7/16
N/R	108-33	108-29	N/R	N/R	N/R	N/R	N/S	N/S	1-3/16	N/R	1-1/2	N/R
N/R	108-33	108-29	N/R	N/R	N/R	N/R	N/S	N/S	1-3/16	N/R	1-1/2	N/R
Yellow	108-33	108-29	108-33	108-29	N/R	134-108	1034-4652	N/S	1-3/8	1-7/16	1-11/16	1-11/16
Plain	108-33	108-29	108-30	108-30	108-27	N/S	134-105	N/S	1-3/32	1-3/32	1-1/2	1-1/2

NUMERICAL LISTING AND COMPONENT PARTS

CARBURETOR PART NO.	CARB. MODEL NO.	CFM	RENEW KIT	TRICK KIT	PRIMARY & SECONDARY NEEDLE & SEAT	PRIMARY MAIN JET	SECONDARY MAIN JET OR PLATE	PRIMARY METERING BLOCK	SECONDARY METERING BLOCK	PRIMARY POWER VALVE	PRIMARY DISCHARGE NOZZLE SIZE
R6425	2300	650	37-656	37-903	6-504	122-82	N/R	134-176	N/S	125-65	.031
R6464	4500	1050	37-487	37-920	6-504	122-88	122-88	134-177	N/R	N/R	.035
R6468-1	4165	650	37-605	37-918	(16,17)	122-60	122-83	134-178	134-167	125-85	.025
R6468-2	4165	650	37-605	37-918	(16,17)	122-602	122-83	134-178	N/A	125-85	.025
R6497	4165	650	37-605	37-918	(16,17)	122-582	122-602	134-179	134-169	(14,15)	.025
R6498	4165	650	37-605	37-918	(16,17)	122-592	122-602	134-179	134-169	(14,15)	.025
R6499	4165	650	37-606	37-919	6-504	122-60	122-63	134-180	134-169	(14,15)	.025
R6512	4165	650	37-605	37-918	(16,17)	122-60	122-60	134-182	134-181	(14,15)	.025
R6520	4160	600	37-119	N/A	6-506	122-62	134-39	N/S	N/S	125-85	.031
R6528	4165	650	37-605	37-918	(16,17)	122-61	122-60	134-183	134-167	(14,15)	.025
R6619-1	4160	600	37-720	37-912	6-506	122-642	134-39	134-184	134-39	125-65	.031
R6647	4150	600	3-655	N/A	6-504	122-68	122-70	N/S	N/S	125-85(12)	.025
R6708	4150	650	37-1272	37-925	6-504	122-552	122-752	134-185	134-39	(21,22)	.025
R6708-1	4150	650	37-1272	37-925	6-504	122-542	122-85	134-185	134-186	125-65	.025
R6709	4150	750	37-1272	37-925	6-504	122-652	122-76	134-182	134-39	(21,22)	.025
R6710	4165	800	37-606	37-919	6-504	122-63	122-86	134-191	134-190	(21,22)	.025
R6711	4165	650	37-605	37-918	(16,17)	122-602	122-632	134-179	134-192	(21,22)	.025
R6772	4165	650	37-605	37-918	(16,17)	122-592	122-602	134-179	134-193	(14,15)	.025
R6773	4165	650	37-605	37-918	(16,17)	122-592	122-602	134-179	134-169	(14,15)	.025
R6774	4165	650	37-605	37-918	(16,17)	122-572	122-602	134-179	134-167	(14,15)	.025
R6846	2300	300	N/A	N/A	6-511	122-60	N/R	N/S	N/R	125-50	.028
R6853	4165	650	37-605	37-918	(16,17)	122-60	122-62	134-166	134-167	(14,15)	.025
R6895	4150	390	37-739	N/A	6-504	122-50	122-62	134-194	134-195	125-85	.025
R6909	4160	600	37-119	37-912	6-506	122-622	134-39	134-184	134-39	125-65	.031
R6910	4165	800	37-606	37-919	6-504	122-612	122-86	134-168	134-192	(14,15)	.025
R6919	4160	600	37-1415	37-912	6-506	122-622	134-39	134-184	134-39	125-206	.031
R6946-1	4160	600	3-1012	N/A	6-504	122-612	134-41	N/S	N/S	125-211	.025
R6947	4160	600	3-1012	N/A	6-504	122-612	134-41	N/S	N/S	125-206	.025
R6979	4160	600	37-747	37-912	6-506	122-642	134-39	134-184	134-39	125-85	.031
R6979-1	4160	600	37-747	37-912	6-506	122-642	134-39	134-184	134-39	125-208	.031
R6989	4160	600	37-1415	37-912	6-506	122-622	134-39	134-184	134-39	125-206	.031
R7001	4165	650	37-743	37-918	(16,17)	122-582	122-602	134-182	134-167	(15,24)	.025
R7002-1	4175	650	37-732	37-922	(16,17)	122-582	134-21	134-179	134-21	125-85	.025
R7004-1	4175	650	37-741	37-922	(16,17)	122-562	134-45	134-187	134-45	125-212	.025
R7004-2	4175	650	37-741	37-922	(16,17)	122-542	134-50	134-198	134-50	125-211	.025
R7005-1	4175	650	37-741	37-922	(16,17)	122-562	134-45	134-187	134-45	125-212	.025
R7005-2	4175	650	37-741	37-922	(16,17)	122-542	134-50	134-198	134-50	125-212	.025
R7006-1	4175	650	37-741	37-922	(16,17)	122-562	134-45	134-187	134-45	125-212	.025
R7006-2	4175	650	37-741	37-922	(16,17)	122-542	134-50	134-198	134-50	125-211	.025
R7009-1	4160	600	37-1415	37-912	6-506	122-622	134-39	134-184	134-39	125-206	.031
R7010	4160	780	37-740	37-925	6-506	122-662	134-42	134-199	134-42	125-65	.025
R7036	2300	300	703-32	N/A	6-511	122-60	N/R	N/S	N/R	125-50	.028
R7053-1	4160	600	37-119	N/A	6-506	122-632	134-39	N/S	N/S	125-85	.031
R7054	4165	650	37-605	37-918	(16,17)	122-592	122-602	134-179	134-169	(14,15)	.025
R7128	4160	650	703-33	N/A	6-511	122-73	134-47	N/S	N/R	125-65	.026
R7154	4160	600	37-119	N/A	6-506	122-62	134-43	N/S	N/S	125-85	.031
R7159	4160	450	703-33	N/A	6-511	122-59	134-8	N/S	N/R	125-85	.021
R7163	4160	600	703-33	N/A	6-511	122-66	134-36	N/S	N/R	125-25	.025
R7320	4500	1150	37-487	37-920	6-504	122-95	122-95	134-200	134-200	N/A	.031
R7320-1	4500	1150	37-487	37-920	6-518	122-95	122-95	134-200	134-200	N/A	.031
R7343	5200	230	37-716	N/A	6-512	124-131	124-139	N/S	N/S	125-36	.020
R7344	5210	255	37-687	N/A	6-512	124-131	124-143	N/S	N/S	125-36	.021
R7351	4175	650	37-741	37-922	(16,17)	122-592	134-22	134-201	134-21	125-206	.037
R7397	4175	650	37-741	37-922	(16,17)	122-582	134-21	134-202	134-21	125-206	.037
R7410	4150	340	37-739	N/A	6-504	122-50	122-62	N/S	N/S	125-85	.025
R7411	4150	370	37-739	N/A	6-504	122-50	122-62	N/S	N/S	125-85	.025
R7413	4160	600	37-119	N/A	6-506	122-632	134-39	N/S	N/S	125-85	.031
R7448	2300	350	37-749	37-901	6-504	122-61	N/A	134-203	N/R	125-85	.031
R7454	4360	450	37-750	N/A	6-514	124-215	124-550	N/R	N/R	125-36	.028
R7455	4360	450	37-750	N/A	6-514	124-215	124-537	N/R	N/R	125-36	.028
R7456	4360	450	37-750	N/A	6-514	124-215	124-550	N/R	N/R	125-36	.028
R7555	4360	450	37-750	N/A	6-514	124-215	124-550	N/R	N/R	125-36	.028
R7556	4360	450	37-750	N/A	6-514	124-215	124-550	N/R	N/R	125-36	.028

NUMERICAL LISTING AND COMPONENT PARTS

SECONDARY NOZZLE SIZE OR SPRING COLOR	PRIMARY BOWL GASKET†	PRIMARY METERING BLOCK GASKET†	SECONDARY BOWL GASKET†	SECONDARY METERING BLOCK GASKET†	SECONDARY METERING PLATE GASKET†	PRIMARY FUEL BOWL	SECONDARY FUEL BOWL	THROTTLE BODY & SHAFT ASSEMBLY	VENTURI DIAMETER PRIMARY	VENTURI DIAMETER SECONDARY	THROTTLE BORE DIAMETER PRIMARY	THROTTLE BORE DIAMETER SECONDARY
N/R	108-32	108-35	N/R	N/R	N/R	134-111	N/S	112-49	1-7/16	N/R	1-3/4	N/R
.035	108-33	108-36	108-33	108-36	N/R	134-108	134-112	N/S	1-11/16	1-11/16	2	2
.037	108-32	108-31	108-32	108-31	N/R	134-110	134-114	112-50	1-5/32	1-3/8	1-3/8	2
.037	108-32	108-31	108-32	108-31	N/R	134-110	134-114	112-51	1-5/32	1-3/8	1-3/8	2
.037	108-32	108-31	108-32	108-31	N/R	134-110	134-114	112-52	1-5/32	1-3/8	1-3/8	2
.037	108-32	108-31	108-32	108-31	N/R	134-110	134-114	112-53	1-5/32	1-3/8	1-3/8	2
.037	108-32	108-31	108-32	108-31	N/R	134-111	134-113	112-50	1-5/32	1-3/8	1-3/8	2
.037	108-32	108-31	108-32	108-31	N/R	134-110	134-114	112-54	1-5/32	1-3/8	1-3/8	2
Purple	108-33	108-31	108-30	108-30	108-27	N/S	N/S	N/S	1-1/4	1-5/16	1-9/16	1-9/16
.037	108-32	108-31	108-32	108-31	N/R	134-110	134-114	112-55	1-5/32	1-3/8	1-3/8	2
Black	108-33	108-31	108-30	108-30	108-27	134-122	134-105	112-11	1-1/4	1-5/16	1-9/16	1-9/16
Yellow	108-33	108-29	108-33	108-29	N/R	N/S	N/S	N/S	1-1/4	1-5/16	1-916	1-9/16
.037	108-32	108-31	108-32	108-31	N/R	134-116	134-117	112-30	1-3/32	1-9/16	1-1/2	1-3/4
.037	108-32	108-31	108-32	108-31	N/R	134-116	134-117	112-30	1-3/32	1-9/16	1-1/2	1-3/4
.037	108-32	108-31	108-32	108-31	N/R	134-116	134-117	112-31	1-1/4	1-9/16	1-1/2	1-3/4
.037	108-32	108-31	108-32	108-31	N/R	134-116	134-117	112-56	1-5/32	1-23/32	1-3/8	2
.028	108-32	108-31	108-32	108-31	N/R	134-115	134-114	112-57	1-5/32	1-3/8	1-3/8	2
.040	108-32	108-31	108-32	108-31	N/R	134-110	134-114	112-58	1-5/32	1-3/8	1-3/8	2
.040	108-32	108-31	108-32	108-31	N/R	134-110	134-114	112-59	1-5/32	1-3/8	1-3/8	2
.037	108-32	108-31	108-32	108-31	N/R	134-110	134-114	112-60	1-5/32	1-3/8	1-3/8	2
N/R	108-33	108-31	N/R	N/R	N/R	N/S	N/R	N/S	1-3/16	N/R	1-1/2	N/R
.037	108-32	108-31	108-32	108-31	N/R	134-110	134-114	112-45	1-5/32	1-3/8	1-3/8	2
.025	108-33	108-29	108-33	108-29	N/R	134-103	134-104	112-23	1-1/16	1-1/16	1-7/16	1-7/16
Black	108-33	108-31	108-30	108-30	108-27	134-120	134-105	112-61	1-1/4	1-5/16	1-9/16	1-9/16
.037	108-32	108-31	108-32	108-31	N/R	134-121	134-117	112-62	1-5/32	1-23/32	1-3/8	2
Black	108-33	108-31	108-30	108-30	108-27	134-122	134-105	112-12	1-1/4	1-5/16	1-9/16	1-9/16
Plain	108-33	108-31	108-30	108-30	108-27	N/S	N/S	1-3/16	1-1/4	1-1/2	1-1/2	
Plain	108-33	108-31	108-30	108-30	108-27	N/S	N/S	1-3/16	1-1/4	1-1/2	1-1/2	
Black	108-33	108-31	108-30	108-30	108-27	134-101	134-105	N/S	1-1/4	1-5/16	1-9/16	1-9/16
Black	108-33	108-31	108-30	108-30	108-27	134-101	134-105	112-63	1-1/4	1-5/16	1-9/16	1-9/16
Black	108-33	108-31	108-30	108-30	108-27	134-122	134-105	112-64	1-1/4	1-5/16	1-9/16	1-9/16
.037	108-32	108-31	108-32	108-31	N/R	134-110	134-114	112-65	1-5/32	1-3/8	1-3/8	2
Black	108-32	108-31	108-30	108-30	108-27	134-110	134-119	112-67	1-5/32	1-3/8	1-3/8	2
Plain	108-32	108-31	108-30	108-30	108-27	134-115	134-119	112-68	1-5/32	1-3/8	1-3/8	2
Plain	108-32	108-31	108-30	108-30	108-27	134-115	134-119	112-69	1-5/32	1-3/8	1-3/8	2
Plain	108-32	108-31	108-30	108-30	108-27	134-115	134-119	112-68	1-5/32	1-3/8	1-3/8	2
Plain	108-32	108-31	108-30	108-30	108-27	134-115	134-119	112-69	1-5/32	1-3/8	1-3/8	2
Plain	108-32	108-31	108-30	108-30	108-27	134-115	134-119	N/S	1-5/32	1-3/8	1-3/8	2
Plain	108-32	108-31	108-30	108-30	108-27	134-115	134-119	112-70	1-5/32	1-3/8	1-3/8	2
Black	108-33	108-31	108-30	108-30	108-27	134-120	134-105	112-72	1-1/4	1-5/16	1-9/16	1-9/16
Black	108-32	108-31	108-30	108-30	108-27	134-116	134-102	112-73	1-1/4	1-9/16	1-1/2	1-3/4
N/R	108-33	108-31	N/R	N/R	N/R	N/S	N/R	N/S	1-3/16	N/R	1-1/2	N/R
Purple	108-33	108-31	108-30	108-30	108-27	N/S	N/S	N/S	1-1/4	1-5/16	1-9/16	1-9/16
.037	108-32	108-31	108-33	108-31	N/R	134-110	134-114	112-74	1-5/32	1-3/8	1-3/8	2
Yellow	108-33	108-31	108-30	108-30	108-27	N/S	N/S	N/S	1-1/4	1-5/16	1-9/16	1-9/16
Purple	108-33	108-31	108-30	108-30	108-27	N/S	N/S	N/S	1-1/4	1-5/16	1-9/16	1-9/16
Plain	108-33	108-31	108-30	108-30	108-27	N/S	N/S	1-3/32	1-3/32	1-1/2	1-1/2	
Purple	108-33	108-31	108-30	108-30	108-27	N/S	N/S	N/S	1-1/4	1-5/16	1-9/16	1-9/16
.035	108-33	108-29	108-33	108-29	N/R	134-108	134-112	N/S	1-13/16	1-13/16	2	2
.035	108-33	108-29	108-33	108-29	N/R	134-108	134-112	N/S	1-13/16	1-13/16	2	2
N/R	N/R	N/R	N/R	N/R	N/R	N/R	N/R	N/R	1-1/25	1-1/16	1-7/25	1-7/16
N/R	N/R	N/R	N/R	N/R	N/R	N/R	N/R	N/R	1-1/33	1-1/16	1-1/4	1-7/25
Black	108-32	108-31	108-30	108-30	108-27	134-110	134-119	112-75	1-13/64	1-13/32	1-3/8	2
Black	108-32	108-31	108-30	108-30	108-27	134-110	134-119	112-76	1-13/64	1-13/32	1-3/8	2
.025	108-33	108-29	108-33	108-29	N/R	N/S	N/S	N/S	1-1/16	1-1/16	1-7/16	1-7/16
.025	108-33	108-29	108-33	108-29	N/R	N/S	N/S	N/S	1-1/16	1-1/16	1-7/16	1-7/16
Purple	108-33	108-31	108-30	108-30	108-27	N/S	N/S	N/S	1-1/4	1-5/16	1-9/16	1-9/16
N/R	108-33	108-29	N/R	N/R	N/R	134-103	N/R	112-14	1-3/16	N/R	1-1/2	N/R
N/R	108-26(5)	N/R	N/R	N/R	N/R	N/R	N/R	N/S	1-1/16	1-3/16	1-3/8	1-7/16
N/R	108-26(5)	N/R	N/R	N/R	N/R	N/R	N/R	N/S	1-1/16	1-3/16	1-3/8	1-7/16
N/R	108-26(5)	N/R	N/R	N/R	N/R	N/R	N/R	N/S	1-1/16	1-3/16	1-3/8	1-7/16
N/R	108-26(5)	N/R	N/R	N/R	N/R	N/R	N/R	N/S	1-1/16	1-3/16	1-3/8	1-7/16
N/R	108-26(5)	N/R	N/R	N/R	N/R	N/R	N/R	N/S	1-1/16	1-3/16	1-3/8	1-7/16

Super Tuning and Modifying Holley Carburetors

NUMERICAL LISTING AND COMPONENT PARTS

CARBURETOR PART NO.	CARB. MODEL NO.	CFM	RENEW KIT	TRICK KIT	PRIMARY & SECONDARY NEEDLE & SEAT	PRIMARY MAIN JET	SECONDARY MAIN JET OR PLATE	PRIMARY METERING BLOCK	SECONDARY METERING BLOCK	PRIMARY POWER VALVE	PRIMARY DISCHARGE NOZZLE SIZE
R7850	4160	600	37-830	N/A	6-506	122-622	134-39	N/S	N/S	125-85	.031
R7855	4175	650	37-741	N/A	(16,17)	122-562	134-45	134-187	134-45	125-212	.028
R7955	4360	450	37-1138	N/A	6-514	124-219	124-550	N/R	N/R	125-201	.028
R7956	4360	450	37-1138	N/A	6-514	124-239	124-550	N/R	N/R	125-20	.028
R7957	4360	450	37-1138	N/A	6-514	124-219	124-550	N/R	N/R	125-201	.028
R7958	4360	450	37-1138	N/A	6-514	124-219	124-550	N/R	N/R	125-201	.028
R7985	4160	600	37-1415	37-912	6-506	122-632	134-39	134-184	134-39	125-208	.031
R7986	4160	600	37-1415	37-912	6-506	125-652	134-39	134-204	134-39	125-208	.031
R7987	4160	600	37-1415	37-912	6-506	122-612	134-39	134-205	134-39	125-208	.031
R8001	4360	450	37-1138	N/A	6-514	124-215	124-550	N/R	N/R	125-201	.028
R8002	4360	450	37-1138	N/A	6-514	124-215	124-550	N/R	N/R	125-201	.028
R8003	4360	450	37-1138	N/A	6-514	124-235	124-550	N/R	N/R	125-201	.028
R8004	4160	600	37-1415	37-912	6-506	122-632	134-39	134-206	134-39	125-208	.031
R8005	4160	600	37-1415	37-912	6-506	122-622	134-39	134-184	134-39	125-208	.031
R8006	4160	600	37-1415	37-912	6-506	122-622	134-39	134-207	134-39	125-208	.031
R8007	4160	390	37-720	37-912	6-506	122-51	134-34	134-208	134-34	125-65	.025
R8059	4175	650	37-741	N/A	(16,17)	122-582	134-21	134-202	134-21	125-206	.037
R8059-1	4175	650	37-741	N/A	(16,17)	122-582	134-49	134-210	134-49	125-211	.025
R8060	4175	650	37-741	N/A	(16,17)	122-582	134-21	134-202	134-21	125-206	.037
R8060-1	4175	650	37-741	N/A	(16,17)	122-582	134-49	134-210	134-49	125-24	.025
R8082	4500	1050	37-487	37-920	6-504	122-84	122-84	134-211	134-200	125-65	.035
R8082-1	4500	1050	37-487	37-920	6-504	122-88	122-88	134-211	134-200	125-65(15)	.035
R8082-2	4500	1050	37-487	37-920	6-518	122-84	122-84	134-211	134-200	125-65(15)	.035
R8123	4160	600	703-22	N/A	6-511	122-66	134-41	N/S	N/R	125-50	.025
R8149	4360	450	37-1138	N/A	6-514	124-231	124-550	N/R	N/R	125-36	.028
R8149-1	4360	450	37-1138	N/A	6-514	124-215	124-550	N/R	N/R	125-201	.028
R8156	4150	750	37-485	37-910	6-504	122-70	122-83	134-155	134-153	125-65	.028
R8158	4360	450	37-750	N/A	6-514	124-219	124-550	N/R	N/R	125-201	.028
R8159	4160	450	703-33	N/A	6-511	122-59	134-32	N/S	N/R	125-85	.021
R8162	4150	850	37-485	37-916	6-504	122-80	122-80	134-163	134-212	125-65	.031
R8181	4160	600	37-1415	N/A	6-504	122-80	122-80	134-184	134-39	125-65(15)	.031
R8203	4360	450	37-1138	N/A	6-514	124-211	124-550	N/R	N/R	125-201	.028
R8204	4360	450	37-1138	N/A	6-514	124-215	124-550	N/R	N/R	125-201	.028
R8206	4360	450	37-1138	N/A	6-514	124-203	124-550	N/R	N/R	125-201	.028
R8207	4160	600	37-830	N/A	6-506	122-622	134-39	N/S	N/S	125-85	.031
R8276	4175	650	37-732	37-922	(16,17)	122-572	134-21	N/S	N/S	125-85	.025
R8302	4175	650	37-732	37-922	(16,17)	122-582	134-21	N/S	N/S	125-85	.025
R8479	4360	450	37-1138	N/A	6-514	124-219	124-589	N/R	N/R	125-201	.028
R8516	4360	450	37-1138	N/A	6-514	124-167	124-423	N/R	N/R	125-203	.028
R8517	4360	450	37-1138	N/A	6-514	124-203	124-524	N/R	N/R	125-201	.028
R8546	4175	650	37-732	N/A	(16,17)	122-582	134-21	134-179	134-21	125-85	.025
R8572	4150	715	703-2	N/A	6-504	122-72	122-84	N/S	N/S	125-85	.026
R8642	4360	450	37-1138	N/A	6-514	124-215	124-550	N/R	N/R	125-201	.028
R8677	4360	450	37-1138	N/A	6-514	124-219	124-524	N/R	N/R	125-200	.028
R8679	4175	650	37-732	37-922	(16,17)	122-592	134-27	134-179	134-27	125-85	.025
R8700	4175	650	37-732	37-922	(16,17)	122-582	134-21	134-179	134-21	125-85	.025
R8771	4360	450	37-1138	N/A	6-514	124-207	124-537	N/R	N/R	125-201	.028
R8804	4150	830	37-485	37-916	6-504	122-80	122-80	134-212	134-212	125-65(B)	.028
R8874	4360	450	37-1138	N/A	6-514	124-219	124-589	N/R	N/R	125-201	.028
R8875	4360	450	37-1180	N/A	6-514	124-231	124-576	N/R	N/R	125-204	.028
R8876	4360	450	37-1179	N/A	6-514	124-231	124-550	N/R	N/R	125-205	.028
R8877	4360	450	37-1160	N/A	6-514	124-231	124-550	N/R	N/R	125-202	.028
R8879	4175	650	37-732	37-922	(16,17)	122-592	134-21	134-179	134-21	125-65	.025
R8896	4500	1050	37-487	37-920	6-504	122-88	122-88	134-213	134-213	N/R	.035
R8896-1	4500	1050	37-487	37-920	6-518	122-88	122-88	134-263	134-263	N/R	.035
R8914	4360	450	37-1138	N/A	6-514	124-207	124-537	N/R	N/R	125-201	.028
R8958	4360	450	37-1138	N/A	6-514	124-195	124-550	N/R	N/R	125-200	.028
R9002	4160	600	37-1415	N/A	6-506	122-632	134-37	134-214	134-37	125-208	.031
R9011	2300	500	37-474	N/A	6-504	122-75	N/R	N/S	N/R	125-50	.028
R9013	4160	600	3-720	N/A	6-506	122-64	134-44	N/S	N/R	125-65	.031
R9015	4160	750	3-720	N/A	6-504	122-76	134-27	N/S	N/R	125-105	.025
R9022	4150	800	703-10	N/A	6-504	122-72	122-87	N/S	N/S	125-65	.031
R9023	4165	800	703-19	N/A	6-504	122-61	122-86	N/S	N/S	125-85(15)	.025

NUMERICAL LISTING AND COMPONENT PARTS

SECONDARY NOZZLE SIZE OR SPRING COLOR	PRIMARY BOWL GASKET†	PRIMARY METERING BLOCK GASKET†	SECONDARY BOWL GASKET†	SECONDARY METERING BLOCK GASKET†	SECONDARY METERING PLATE GASKET†	PRIMARY FUEL BOWL	SECONDARY FUEL BOWL	THROTTLE BODY & SHAFT ASSEMBLY	VENTURI DIAMETER PRIMARY	VENTURI DIAMETER SECONDARY	THROTTLE BORE DIAMETER PRIMARY	THROTTLE BORE DIAMETER SECONDARY
Plain	108-33	108-31	108-30	108-30	108-27	N/S	N/S	N/S	1-1/4	1-5/16	1-9/16	1-9/16
Plain	108-32	108-31	108-30	108-30	108-27	134-115	134-119	112-77	1-13/32	1-13/64	1-3/8	2
N/R	108-26(5)	N/R	N/R	N/R	N/R	N/R	N/R	N/S	1-1/16	1-3/16	1-3/8	1-7/16
N/R	108-26(5)	N/R	N/R	N/R	N/R	N/R	N/R	N/S	1-1/16	1-3/16	1-3/8	1-7/16
N/R	108-26(5)	N/R	N/R	N/R	N/R	N/R	N/R	N/S	1-1/16	1-3/16	1-3/8	1-7/16
Black	108-33	108-31	108-30	108-30	108-27	134-101	134-105	112-78	1-1/4	1-5/16	1-9/16	1-9/16
Black	108-33	108-31	108-30	108-30	108-27	134-124	134-105	112-79	1-1/4	1-5/16	1-9/16	1-9/16
Black	108-33	108-31	108-31	108-30	108-27	134-120	134-105	112-80	1-1/4	1-5/16	1-9/16	1-9/16
N/R	108-26(5)	N/R	N/R	N/R	N/R	N/R	N/R	N/S	1-1/16	1-3/16	1-3/8	1-7/16
N/R	108-26(5)	N/R	N/R	N/R	N/R	N/R	N/R	N/S	1-1/16	1-3/16	1-3/8	1-7/16
Black	108-33	108-31	108-30	108-30	108-27	134-120	134-105	112-81	1-1/4	1-5/16	1-9/16	1-9/16
Black	108-33	108-31	108-30	108-30	108-27	134-124	134-105	112-82	1-1/4	1-5/16	1-9/16	1-9/16
Black	108-33	108-31	108-30	108-30	108-27	134-120	134-105	112-83	1-1/4	1-5/16	1-9/16	1-9/16
Plain	108-33	108-31	108-30	108-30	108-27	134-122	134-105	112-84	1-1/16	1-1/16	1-7/16	1-7/16
Black	108-32	108-31	108-30	108-30	108-27	134-110	134-119	N/S	1-13/64	1-13/32	1-3/8	2
Black	108-32	108-31	108-30	108-30	108-27	134-110	134-119	112-85	1-13/64	1-13/32	1-3/8	2
Black	108-32	108-31	108-30	108-30	108-27	134-110	134-119	N/S	1-13/64	1-13/32	1-3/8	2
Black	108-32	108-31	108-30	108-30	108-27	134-110	134-119	112-86	1-13/64	1-13/32	1-3/8	2
.035	108-33	108-29	108-33	108-29	N/R	134-108	134-112	N/S	1-11/16	1-11/16	2	2
.035	108-33	108-29	108-33	108-29	N/R	134-108	134-112	N/S	1-11/16	1-11/16	2	2
.035	108-33	108-29	108-33	108-29	N/R	134-108	134-112	N/S	1-11/16	1-11/16	2	2
Purple	108-33	108-31	108-30	108-30	108-27	N/S	N/S	N/S	1-1/4	1-5/16	1-9/16	1-9/16
N/R	108-26(5)	N/R	N/R	N/R	N/R	N/R	N/R	N/S	1-1/16	1-3/16	1-3/8	1-7/16
N/R	108-26(5)	N/R	N/R	N/R	N/R	N/R	N/R	N/S	1-1/16	1-3/16	1-3/8	1-7/16
.031	108-33	108-29	108-33	108-29	N/R	134-103	134-104	112-32	1-3/8	1-3/8	1-11/16	1-11/16
N/R	108-26(5)	N/R	N/R	N/R	N/R	N/R	N/R	N/S	1-1/16	1-3/16	1-3/8	1-7/16
Plain	108-33	108-31	108-30	108-30	108-27	N/S	N/S	N/S	1-3/32	1-3/32	1-1/2	1-1/2
.031	108-33	108-29	108-33	108-29	N/R	134-103	134-104	112-33	1-9/16	1-9/16	1-3/4	1-3/4
.031	108-33	108-29	108-33	108-29	N/R	134-122	134-105	112-64	1-9/16	1-9/16	1-3/4	1-3/4
N/R	108-26(5)	N/R	N/R	N/R	N/R	N/R	N/R	N/S	1-1/16	1-3/16	1-3/8	1-7/16
N/R	108-26(5)	N/R	N/R	N/R	N/R	N/R	N/R	N/S	1-1/16	1-3/16	1-3/8	1-7/16
Plain	108-33	108-31	108-30	108-30	108-27	N/S	N/S	N/S	1-1/4	1-5/16	1-9/16	1-9/16
Black	108-32	108-31	108-30	108-30	108-27	N/S	N/S	N/S	1-13/64	1-13/32	1-3/8	2
Black	108-32	108-31	108-30	108-30	108-27	N/S	N/S	N/S	1-13/64	1-13/32	1-3/8	2
N/R	108-26(5)	N/R	N/R	N/R	N/R	N/R	N/R	N/S	1-1/16	1-3/16	1-3/8	1-7/16
N/R	108-26(5)	N/R	N/R	N/R	N/R	N/R	N/R	N/S	1-1/16	1-3/16	1-3/8	1-7/16
Black	108-32	108-31	108-30	108-30	108-27	134-110	134-119	112-87	1-13/64	1-13/32	1-3/8	2
Yellow	108-33	108-29	108-33	108-29	N/R	134-108	1034-4652	N/S	1-3/8	1-7/16	1-11/16	1-11/16
N/R	108-26(5)	N/R	N/R	N/R	N/R	N/R	N/R	N/S	1-1/16	1-3/16	1-3/8	1-7/16
N/R	108-26(5)	N/R	N/R	N/R	N/R	N/R	N/R	N/S	1-1/16	1-3/16	1-3/8	1-7/16
Plain	108-32	108-31	108-30	108-30	108-27	134-110	134-119	112-88	1-13/64	1-13/32	1-3/8	2
Black	108-32	108-31	108-30	108-30	108-27	134-110	134-119	112-87	1-13/64	1-13/32	1-3/8	2
N/R	108-26(5)	N/R	N/R	N/R	N/R	N/R	N/R	N/S	1-1/16	1-3/16	1-3/8	1-7/16
.028	108-33	108-29	108-33	108-29	N/R	134-103	134-104	112-34	1-9/16	1-9/16	1-11/16	1-11/16
N/R	108-26(5)	N/R	N/R	N/R	N/R	N/R	N/R	N/S	1-1/16	1-3/16	1-3/8	1-7/16
N/R	108-26(5)	N/R	N/R	N/R	N/R	N/R	N/R	N/S	1-1/16	1-3/16	1-3/8	1-7/16
N/R	108-26(5)	N/R	N/R	N/R	N/R	N/R	N/R	N/S	1-1/16	1-3/16	1-3/8	1-7/16
N/R	108-26(5)	N/R	N/R	N/R	N/R	N/R	N/R	N/S	1-1/16	1-3/16	1-3/8	1-7/16
Black	108-32	108-31	108-30	108-30	108-27	134-110	134-119	112-89	1-13/64	1-13/32	1-3/8	2
.035	108-33	108-36	108-33	108-36	N/R	134-108	134-112	N/S	1-11/16	1-11/16	2	2
.037	108-81	108-36	108-81	108-36	N/R	134-108	134-112	N/S	1-11/16	1-11/16	2	2
N/R	108-26(5)	N/R	N/R	N/R	N/R	N/R	N/R	N/S	1-1/16	1-3/16	1-3/8	1-7/16
N/R	108-26(5)	N/R	N/R	N/R	N/R	N/R	N/R	N/S	1-1/16	1-3/16	1-3/8	1-7/16
Black	108-33	108-31	108-30	108-30	108-27	134-122	134-105	N/S	1-1/4	1-5/16	1-9/16	1-9/16
N/R	108-33	108-29	N/R	N/R	N/R	N/S	N/R	N/S	1-3/8	N/R	1-11/16	N/R
Black	108-33	108-31	108-30	108-30	108-27	N/S	N/S	N/S	1-1/4	1-5/16	1-9/16	1-9/16
Plain	108-33	108-29	108-30	108-30	108-27	N/S	N/S	N/S	1-3/8	1-7/16	1-11/16	1-11/16
.031	108-33	108-29	108-33	108-29	N/R	N/S	N/S	N/S	1-3/8	1-7/16	1-11/16	1-11/16
.037	108-32	108-31	108-32	108-31	N/R	N/S	N/S	N/S	1-5/32	1-23/32	1-3/8	2

Super Tuning and Modifying Holley Carburetors

NUMERICAL LISTING AND COMPONENT PARTS

CARBURETOR PART NO.	CARB. MODEL NO.	CFM	RENEW KIT	TRICK KIT	PRIMARY & SECONDARY NEEDLE & SEAT	PRIMARY MAIN JET	SECONDARY MAIN JET OR PLATE	PRIMARY METERING BLOCK	SECONDARY METERING BLOCK	PRIMARY POWER VALVE	PRIMARY DISCHARGE NOZZLE SIZE
R9029	4150	715	703-2	N/A	6-504	122-75	122-84	N/S	N/S	125-85	.026
R9040	4160	600	37-1177	N/A	6-85	122-661	134-23	134-215	134-23	125-211	.031
R9088	4360	450	37-1316	N/A	6-514	124-215	124-550	N/R	N/R	125-214	.028
R9105	4360	450	37-1160	N/A	6-514	124-195	124-550	N/R	N/R	125-200	.028
R9112	4360	450	37-1138	N/A	6-514	124-211	124-563	N/R	N/R	125-201	.028
R9162	4360	450	37-1138	N/A	6-514	124-203	124-537	N/R	N/R	125-201	.028
R9185	4360	450	37-1138	N/A	6-514	124-191	124-550	N/R	N/R	125-201	.028
R9188	4150	780	37-1184	37-915	6-504	122-72	122-76	N/S	N/S	(12,21)	.025
R9192	4360	450	37-1138	N/A	6-514	124-231	124-550	N/R	N/R	125-201	.028
R9193	4360	450	37-1138	N/A	6-514	124-211	124-589	N/R	N/R	125-201	.028
R9210	4160	600	37-1415	N/A	6-506	122-612	134-39	N/S	N/S	125-208	.031
R9219	4160	600	37-1415	N/A	6-506	122-632	134-39	134-184	134-39	125-208	.031
R9228	5200	280	37-1279	N/A	6-512	124-163	124-231	N/R	N/R	125-36	.023
R9254	4160	600	37-1415	N/A	6-506	122-622	134-39	N/S	N/S	125-211	.031
R9375	4500	1050	37-487	37-920	6-504	122-92	122-92	134-213	134-213	N/R	.035
R9375-1	4500	1050	37-487	37-920	6-518	122-88	122-88	134-263	134-263	N/R	.035
R9377	4500	1150	37-487	37-920	6-504	122-94	122-94	134-175	134-175	N/R	.035
R9377-1	4500	1150	37-487	37-920	6-518	122-92	122-92	134-264	134-264	N/R	.035
R9379	4150	750	37-485	37-910	6-504	122-68	122-81	134-155	134-153	125-65	.028
R9380	4150	850	37-485	37-916	6-504	122-78	122-78	134-163	134-212	125-65(15)	.031
R9381	4150	830	37-485	37-916	6-504	122-78	122-78	134-163	134-212	125-65(15)	.028
R9392	4160	600	703-33	N/A	6-511	122-66	N/S	N/S	N/R	125-25	.025
R9393	4160	450	703-28	N/A	6-511	122-59	N/S	N/S	N/R	125-85	.021
R9394	4160	650	703-28	N/A	6-511	122-73	N/S	N/S	N/R	125-65	.026
R9399	4160	650	703-28	N/A	6-511	122-73	N/S	N/S	N/R	125-65	.040
R9399-1	4160	650	703-28	N/A	6-511	122-73	N/S	N/S	N/R	125-65	.040
R9429	5200	280	37-1279	N/A	6-512	124-183	124-231	N/R	N/R	125-36	.023
R9441	5200	280	37-1279	N/A	6-512	124-163	124-231	N/R	N/R	125-36	.023
R9444	5200	280	37-1279	N/A	6-512	124-163	124-231	N/R	N/R	125-36	.023
R9446	5200	280	37-1279	N/A	6-512	124-163	124-231	N/R	N/R	125-36	.023
R9545	5200	280	37-1279	N/A	6-512	124-183	124-231	N/R	N/R	125-36	.023
R9626	4160	600	37-1274	N/A	6-506	122-612	134-39	N/S	N/S	125-206	.031
R9644	6520	280	37-1205	N/A	N/A	124-179	124-283	N/R	N/R	N/A	.020
R9645	4150	750	37-1312	37-923	6-515	122-80	122-80	134-220	134-219	125-65(15)	.045
R9646	4150	850	37-1312	37-923	6-515	122-92	122-92	134-220	134-219	125-65(15)	.045
R9647	2300	500	37-1311	37-924	6-515	122-81	N/R	134-222	N/R	125-45	.040
R9655	6520	280	37-1205	N/A	N/A	124-195	124-299	N/R	N/R	N/A	.020
R9659	6520	280	37-1205	N/A	N/A	124-131	134-267	N/R	N/R	N/A	.020
R9678	4360	450	37-1160	N/A	6-514	124-211	124-550	N/R	N/R	125-202	.028
R9681	5200	280	37-1279	N/A	6-512	124-171	124-215	N/R	N/R	125-31	.023
R9682	6520	280	37-1205	N/A	N/A	124-219	124-283	N/R	N/R	N/A	.020
R9688	5200	280	37-1279	N/A	6-512	124-163	124-251	N/R	N/R	125-36	.023
R9689	5200	280	37-1279	N/A	6-512	124-159	124-251	N/R	N/R	125-36	.023
R9694	4360	450	37-1138	N/A	6-514	124-171	124-485	N/R	N/R	125-201	.028
R9767	5200	280	37-1279	N/A	6-512	124-179	124-259	N/R	N/R	125-36	.023
R9776	4160	450	37-1321	N/A	6-506	122-582	134-6	134-251	134-6	125-85	.031
R9777	4360	450	37-750	N/A	6-514	124-255	124-550	N/R	N/R	125-36	.028
R9781	5200	280	37-1279	N/A	6-512	124-159	124-251	N/R	N/R	125-36	.023
R9810	6520	280	37-1205	N/A	N/A	124-195	124-299	N/R	N/R	N/A	.020
R9811	6520	280	37-1205	N/A	N/A	124-155	124-271	N/R	N/R	N/A	.020
R9834	4160	600	37-720	37-912	6-506	122-642	134-39	134-184	134-39	125-65	.031
R9834-1	4160	600	37-720	37-912	6-506	122-661	134-39	134-241	134-39	125-65	.031
R9834-2	4160	600	37-720	37-912	6-506	122-68	134-39	134-241	134-39	125-65	.031
R9834-3	4160	600	37-720	37-912	6-506	122-68	134-39	134-241	134-39	125-65	.031
R9864	5200	280	37-1279	N/A	6-512	124-159	124-219	N/R	N/R	125-36	.023
R9875	4360	450	37-1316	N/A	6-514	124-199	124-576	N/R	N/R	125-214	.028
R9895	4175	650	37-741	N/A	(16,17)	122-592	134-21	134-201	134-21	125-206	.037
R9896	6510	280	37-1286	N/A	N/A	124-104	124-271	N/R	N/R	N/A	.020
R9899	5200	280	37-1279	N/A	6-512	124-147	124-231	N/R	N/R	125-36	.023
R9923	4175	650	37-741	N/A	(16,17)	122-542	134-50	N/S	N/S	125-211	.025
R9925	5200	280	37-1279	N/A	6-512	124-147	124-251	N/R	N/R	125-36	.023
R9931	4360	450	37-1138	N/A	6-514	124-239	124-550	N/R	N/R	125-203	.028
R9932	5200	280	37-1279	N/A	6-512	124-159	124-219	N/R	N/R	125-36	.023

NUMERICAL LISTING AND COMPONENT PARTS

SECONDARY NOZZLE SIZE OR SPRING COLOR	PRIMARY BOWL GASKET†	PRIMARY METERING BLOCK GASKET†	SECONDARY BOWL GASKET†	SECONDARY METERING BLOCK GASKET†	SECONDARY METERING PLATE GASKET†	PRIMARY FUEL BOWL	SECONDARY FUEL BOWL	THROTTLE BODY & SHAFT ASSEMBLY	VENTURI DIAMETER PRIMARY	VENTURI DIAMETER SECONDARY	THROTTLE BORE DIAMETER PRIMARY	THROTTLE BORE DIAMETER SECONDARY
Yellow	108-33	108-29	108-33	108-29	N/R	N/S	N/S	N/S	1-3/8	1-7/16	1-11/16	1-11/16
Plain	108-33	108-31	108-30	108-30	108-27	134-107	134-105	112-90	1-1/4	1-5/16	1-9/16	1-9/16
N/R	108-26(5)	N/R	N/R	N/R	N/R	N/R	N/R	N/S	1-1/16	1-3/16	1-3/8	1-7/16
N/R	108-26(5)	N/R	N/R	N/R	N/R	N/R	N/R	N/S	1-1/16	1-3/16	1-3/8	1-7/16
N/R	108-26(5)	N/R	N/R	N/R	N/R	N/R	N/R	N/S	1-1/16	1-3/16	1-3/8	1-7/16
N/R	108-26(5)	N/R	N/R	N/R	N/R	N/R	N/R	N/S	1-1/16	1-3/16	1-3/8	1-7/16
N/R	108-26(5)	N/R	N/R	N/R	N/R	N/R	N/R	N/S	1-1/16	1-3/16	1-3/8	1-7/16
Plain	108-33	108-29	108-33	108-29	N/R	N/S	N/S	N/S	1-3/8	1-7/16	1-11/16	1-11/16
N/R	108-26(5)	N/R	N/R	N/R	N/R	N/R	N/R	N/S	1-1/16	1-3/16	1-3/8	1-7/16
N/R	108-26(5)	N/R	N/R	N/R	N/R	N/R	N/R	N/S	1-1/16	1-3/16	1-3/8	1-7/16
Black	108-33	108-31	108-30	108-30	108-27	N/S	N/S	N/S	1-1/4	1-5/16	1-9/16	1-9/16
Black	108-33	108-31	108-30	108-30	108-27	134-124	134-105	112-91	1-1/4	1-5/16	1-9/16	1-9/16
N/R	N/R	N/R	N/R	N/R	N/R	N/R	N/R	N/R	1-1/25	1-1/16	1-7/25	1-7/16
Black	108-33	108-31	108-30	108-30	108-27	N/S	N/S	N/S	1-1/4	1-5/16	1-9/16	1-9/16
.035	108-33	108-36	108-33	108-36	N/R	134-108	134-112	N/S	1-11/16	1-11/16	2	2
.037	108-81	108-36	108-81	108-36	N/R	134-108	134-112	N/S	1-11/16	1-11/16	2	2
.035	108-33	108-36	108-33	108-36	N/R	134-108	134-112	N/S	1-13/16	1-13/16	2	2
.037	108-81	108-36	108-81	108-36	N/R	134-108	134-112	N/S	1-13/16	1-13/16	2	2
.031	108-33	108-29	108-33	108-29	N/R	134-103	134-104	112-32	1-3/8	1-3/8	1-11/16	1-11/16
.031	108-33	108-29	108-33	108-29	N/R	134-103	134-104	112-33	1-9/16	1-9/16	1-3/4	1-3/4
.028	108-33	108-29	108-33	108-29	N/R	134-103	134-104	112-34	1-9/16	1-9/16	1-11/16	1-11/16
Purple	108-33	108-31	108-30	108-30	108-13	N/S	N/S	N/S	1-1/4	1-5/16	1-9/16	1-9/16
Plain	108-33	108-31	108-30	108-30	108-13	N/S	N/S	N/S	1-3/32	1-3/32	1-1/2	1-1/2
Yellow	108-33	108-31	108-30	108-30	108-13	N/S	N/S	N/S	1-1/4	1-5/16	1-9/16	1-9/16
White	108-33	108-32	108-30	108-30	108-13	N/S	N/S	N/S	1-1/4	1-5/16	1-9/16	1-9/16
White	108-33	108-32	108-30	108-30	108-13	N/S	N/S	N/S	1-1/4	1-5/16	1-9/16	1-9/16
N/R	N/R	N/R	N/R	N/R	N/R	N/R	N/R	N/R	1-1/25	1-1/16	1-7/25	1-7/16
N/R	N/R	N/R	N/R	N/R	N/R	N/R	N/R	N/R	1-1/25	1-1/16	1-7/25	1-7/16
N/R	N/R	N/R	N/R	N/R	N/R	N/R	N/R	N/R	1-1/25	1-1/16	1-7/25	1-7/16
N/R	N/R	N/R	N/R	N/R	N/R	N/R	N/R	N/R	1-1/25	1-1/16	1-7/25	1-7/16
N/R	N/R	N/R	N/R	N/R	N/R	N/R	N/R	N/R	1-1/25	1-1/16	1-7/25	1-7/16
Black	108-33	108-31	108-30	108-30	108-27	N/S	N/S	N/S	1-1/4	1-5/16	1-9/16	1-9/16
N/R	N/R	N/R	N/R	N/R	N/R	N/R	N/R	N/R	1-1/25	1-1/16	1-7/25	1-7/16
.045	108-33	108-29	108-33	108-29	N/R	134-103	134-104	112-35	1-3/8	1-3/8	1-11/16	1-11/16
.045	108-33	108-29	108-33	108-29	N/R	134-103	134-104	112-36	1-9/16	1-9/16	1-3/4	1-3/4
N/R	108-33	108-29	N/R	N/R	N/R	134-103	N/S	112-2	1-3/8	N/R	1-11/16	N/R
N/R	N/R	N/R	N/R	N/R	N/R	N/R	N/R	N/R	1-1/25	1-1/16	1-7/25	1-7/16
N/R	108-26	N/R	N/R	N/R	N/R	N/R	N/R	N/S	1-1/16	1-3/16	1-3/8	1-7/16
N/R	N/R	N/R	N/R	N/R	N/R	N/R	N/R	N/R	1-1/25	1-1/16	1-7/25	1-7/16
N/R	N/R	N/R	N/R	N/R	N/R	N/R	N/R	N/R	1-1/25	1-1/16	1-7/25	1-7/16
N/R	N/R	N/R	N/R	N/R	N/R	N/R	N/R	N/R	1-1/25	1-1/16	1-7/25	1-7/16
N/R	108-26	N/R	N/R	N/R	N/R	N/R	N/R	N/R	1-1/16	1-3/16	1-3/8	1-7/16
N/R	N/R	N/R	N/R	N/R	N/R	N/R	N/R	N/R	1-1/25	1-1/16	1-7/25	1-7/16
N/R	108-33	108-29	108-30	108-30	108-27	134-101	134-105	112-98	1-3/32	1-3/32	1-1/2	1-1/2
N/R	108-26(5)	N/R	N/R	N/R	N/R	N/R	N/R	N/R	1-1/16	1-3/16	1-3/8	1-7/16
N/R	N/R	N/R	N/R	N/R	N/R	N/R	N/R	N/R	1-1/25	1-1/16	1-7/25	1-7/16
N/R	N/R	N/R	N/R	N/R	N/R	N/R	N/R	N/R	1-1/25	1-1/16	1-7/25	1-7/16
Black	108-33	108-31	108-30	108-30	108-27	134-122	134-105	112-92	1-1/4	1-5/16	1-9/16	1-9/16
Black	108-33	108-31	108-30	108-30	108-27	134-122	134-105	112-92	1-1/4	1-5/16	1-9/16	1-9/16
Black	108-33	108-31	108-30	108-30	108-27	134-122	134-105	112-92	1-1/4	1-5/16	1-9/16	1-9/16
Black	108-33	108-31	108-30	108-30	108-27	134-122	134-105	112-92	1-1/4	1-5/16	1-9/16	1-9/16
N/R	N/R	N/R	N/R	N/R	N/R	N/R	N/R	N/R	1-1/25	1-1/16	1-7/25	1-7/16
N/R	108-26	N/R	N/R	N/R	N/R	N/R	N/R	N/R	1-1/16	1-3/16	1-3/8	1-7/16
Black	108-32	108-31	108-30	108-30	108-27	134-110	134-119	112-93	1-13/64	1-13/32	1-3/8	2
N/R	N/R	N/R	N/R	N/R	N/R	N/R	N/R	N/R	1-1/25	1-1/16	1-7/25	1-7/16
N/R	N/R	N/R	N/R	N/R	N/R	N/R	N/R	N/R	1-1/25	1-1/16	1-7/25	1-7/16
Black	108-32	108-31	108-30	108-30	108-27	N/S	N/S	N/S	1-13/64	1-13/32	1-3/8	2
N/R	N/R	N/R	N/R	N/R	N/R	N/R	N/R	N/R	1-1/25	1-1/16	1-7/25	1-7/16
N/R	108-26	N/R	N/R	N/R	N/R	N/R	N/R	N/S	1-1/16	1-3/16	1-3/8	1-7/16
N/R	N/R	N/R	N/R	N/R	N/R	N/R	N/R	N/R	1-1/25	1-1/16	1-7/25	1-7/16

NUMERICAL LISTING AND COMPONENT PARTS

CARBURETOR PART NO.	CARB. MODEL NO.	CFM	RENEW KIT	TRICK KIT	PRIMARY & SECONDARY NEEDLE & SEAT	PRIMARY MAIN JET	SECONDARY MAIN JET OR PLATE	PRIMARY METERING BLOCK	SECONDARY METERING BLOCK	PRIMARY POWER VALVE	PRIMARY DISCHARGE NOZZLE SIZE
R9935	4360	450	37-1138	N/A	6-514	124-207	124-589	N/R	N/R	125-203	.028
R9948	4175	650	37-741	N/A	(16,17)	122-563	134-50	N/S	N/S	125-211	.025
R9973	4360	450	37-1138	N/A	6-514	124-171	124-330	N/R	N/R	125-203	.028
R9976	4175	650	37-741	N/A	(16,17)	122-582	134-49	N/S	N/S	125-211	.025
R50399	4160	650	703-28	N/A	6-511	122-73	N/S	N/S	N/R	125-65	.040
R50399-1	4160	650	703-28	N/A	6-516	122-73	N/S	N/S	N/R	125-65	.040
R50405	4160	650	703-28	N/A	6-516	122-74	N/S	N/S	N/R	125-65	.040
R50405-1	4160	650	703-28	N/A	6-516	122-74	N/S	N/S	N/R	125-65	.040
R50417	2300	300	703-30	N/A	6-511	122-60	N/R	N/S	N/R	125-50	.028
R50417-1	2300	300	703-30	N/A	6-516	122-60	N/R	N/S	N/R	125-50	.028
R50418	4160	450	703-28	N/A	6-511	122-59	N/S	N/S	N/R	125-85	.021
R50419	4160	600	703-29	N/A	6-511	122-66	N/S	N/S	N/R	125-25	.025
R50419-1	4160	600	703-29	N/A	6-516	122-65	N/S	N/S	N/R	125-25	.025
R50419-2	4160	600	703-29	N/A	6-516	122-65	N/S	N/S	N/R	125-25	.025
R50461	2300	300	703-30	N/A	6-511	122-60	N/R	N/S	N/R	125-50	.028
R50461-1	2300	300	703-30	N/A	6-516	122-60	N/R	N/S	N/R	125-50	.028
R50462	4160	450	703-28	N/A	6-516	122-59	N/S	N/S	N/R	125-85	.021
R50462-1	4160	450	703-28	N/A	6-516	122-59	N/S	N/S	N/R	125-85	.021
R50463	4160	600	703-29	N/A	6-516	122-65	N/S	N/S	N/R	125-25	.025
R50463-1	4160	600	703-29	N/A	6-516	122-65	N/S	N/S	N/R	125-25	.025
R50464	4160	750	703-33	N/A	6-511	122-74	N/S	N/S	N/R	125-65	.040
R50467	2300	300	703-30	N/A	6-511	122-61	N/R	N/S	N/R	125-50	.028
R50467-1	2300	300	703-30	N/A	6-516	122-61	N/R	N/S	N/R	125-50	.028
R50468	4160	450	703-28	N/A	6-511	122-59	N/S	N/S	N/R	125-85	.021
R50468-1	4160	450	703-28	N/A	6-516	122-59	N/S	N/S	N/R	125-85	.021
R50469	4160	600	703-47	N/A	6-511	122-65	N/S	N/S	N/R	125-25	.025
R50469-1	4160	600	703-47	N/A	6-516	122-65	N/S	N/S	N/R	125-25	.025
R50470	4160	650	703-33	N/A	6-516	122-74	N/S	N/S	N/R	125-65	.040
R50483	4010	600	703-53	N/A	6-504	122-69	122-76	N/R	N/R	125-65	.026
R50483-1	4010	600	703-53	N/A	6-504	122-69	122-77	N/R	N/R	125-65	.035
R75010	4500	1150	703-44	37-920	6-518	122-92	122-92	134-264	134-264	N/R	.035
R75011	4500	1250	703-44	37-920	6-518	122-97	122-97	134-265	134-265	N/R	.035
R80054	5200	280	37-716	N/A	6-512	124-231	124-247	N/R	N/R	125-36	.023
R80055	5200	280	37-716	N/A	6-512	124-231	124-247	N/R	N/R	125-36	.023
R80056	5200	280	37-716	N/A	6-512	124-231	124-247	N/R	N/R	125-36	.023
R80057	5200	280	37-716	N/A	6-512	124-132	124-135	N/R	N/R	125-36	.023
R80073	4175	650	37-1421	N/A	(16,17)	122-642	134-52	N/S	N/S	125-213	.037
R80086	4360	450	37-1316	N/A	6-514	124-199	124-550	N/R	N/R	125-214	.028
R80095	2305	500	37-749	N/A	6-504	122-55	122-73	N/S	N/R	125-85	.035
R80098	4180	600	37-1346	N/A	6-517	122-612	N/S	N/S	N/S	125-215	.028
R80099	4180	600	37-1346	N/A	6-517	122-622	N/S	N/S	N/S	125-218	.028
R80111	4180	600	37-1346	N/A	6-517	122-612	N/S	N/S	N/S	125-216	.028
R80112	4180	600	37-1346	N/A	6-517	122-622	N/S	N/S	N/S	125-217	.028
R80120	2305	350	37-749	N/A	6-504	122-52	122-65	N/S	N/R	125-85	.035
R80128	4175	650	37-741	N/A	6-510	122-582	134-50	N/S	N/S	125-211	.031
R80133	4180	600	37-1346	N/A	6-517	122-611	N/S	N/S	N/S	125-216	.028
R80134	4180	600	37-1346	N/A	6-517	122-612	N/S	N/S	N/S	1025-609-8	.028
R80135	4180	600	37-1346	N/A	6-517	122-612	N/S	N/S	N/S	1025-609-8	.028
R80136	4180	600	37-1346	N/A	6-517	122-612	N/S	N/S	N/S	1025-609-8	.028
R80137	4180	600	37-1346	N/A	6-517	122-612	N/S	N/S	N/S	1025-609-8	.028
R80139	4175	650	37-741	N/A	6-510	122-592	134-21	N/S	N/S	1025-475-13	.037
R80140	4175	650	37-1421	N/A	6-510	122-642	134-52	N/S	N/S	125-213	.037
R80145	4150	600	37-1184	N/A	6-504	122-68	122-70	134-232	134-233	125-65	.031
R80155	4175	650	37-741	N/A	6-510	122-632	134-21	N/S	N/S	1025-475-13	.037
R80159	4150	715	37-1184	N/A	6-504	122-74	122-85	N/S	N/S	125-85	.026
R80163	4180	600	37-1346	N/A	6-517	122-622	N/S	N/S	N/S	1025-609-8	.028
R80164	4180	600	37-1422	N/A	6-517	122-612	N/S	N/S	N/S	1025-609-8	.028
R80165	4180	600	37-1346	N/A	6-517	122-612	N/S	N/S	N/S	1025-609-8	.028
R80166	4180	600	37-1346	N/A	6-517	122-612	N/S	N/S	N/S	1025-609-8	.028
R80169	4175	650	37-741	N/A	6-510	122-543	134-53	N/S	N/S	125-211	.025
R80180	4150	850	703-46	N/A	6-504	122-92	122-92	N/S	N/S	125-65(15)	.028
R80186	4500	750	37-487	37-920	6-504	122-70	122-70	134-237	134-238	125-65(15)	.028
R80186-1	4500	750	37-487	37-920	6-518	122-70	122-70	134-237	134-238	125-65(15)	.028

NUMERICAL LISTING AND COMPONENT PARTS

SECONDARY NOZZLE SIZE OR SPRING COLOR	PRIMARY BOWL GASKET†	PRIMARY METERING BLOCK GASKET†	SECONDARY BOWL GASKET†	SECONDARY METERING BLOCK GASKET†	SECONDARY METERING PLATE GASKET†	PRIMARY FUEL BOWL	SECONDARY FUEL BOWL	THROTTLE BODY & SHAFT ASSEMBLY	VENTURI DIAMETER PRIMARY	VENTURI DIAMETER SECONDARY	THROTTLE BORE DIAMETER PRIMARY	THROTTLE BORE DIAMETER SECONDARY
N/R	108-26	N/R	N/R	N/R	N/R	N/R	N/R	N/S	1-1/16	1-3/16	1-3/8	1-7/16
Black	108-32	108-31	108-30	108-30	108-27	N/S	N/S	N/S	1-13/64	1-13/32	1-3/8	2
N/R	108-26	N/R	N/R	N/R	N/R	N/R	N/R	N/S	1-1/16	1-3/16	1-3/8	1-7/16
Black	108-32	108-31	108-30	108-30	108-27	134-115	134-119	N/S	1-13/64	1-13/32	1-3/8	2
White	108-33	108-31	108-30	108-30	108-13	N/S	N/S	N/S	1-1/4	1-5/16	1-9/16	1-9/16
White	108-33	108-31	108-30	108-30	108-13	N/S	N/S	N/S	1-1/4	1-5/16	1-9/16	1-9/16
Pink	108-33	108-31	108-30	108-30	108-13	N/S	N/S	N/S	1-1/4	1-5/16	1-9/16	1-9/16
Pink	108-33	108-31	108-30	108-30	108-13	N/S	N/S	N/S	1-1/4	1-5/16	1-9/16	1-9/16
N/R	108-33	108-31	N/R	N/R	N/R	N/R	N/R	N/S	1-3/16	N/R	1-1/2	N/R
N/R	108-33	108-31	N/R	N/R	N/R	N/R	N/R	N/S	1-3/16	N/R	1-1/2	N/R
Plain	108-33	108-31	108-30	108-30	108-13	N/S	N/S	N/S	1-3/32	1-3/32	1-1/2	1-1/2
Purple	108-33	108-31	108-30	108-30	108-13	N/S	N/S	N/S	1-1/4	1-5/16	1-9/16	1-9/16
Purple	108-33	108-31	108-30	108-30	108-13	N/S	N/S	N/S	1-1/4	1-5/16	1-9/16	1-9/16
Purple	108-33	108-31	108-30	108-30	108-13	N/S	N/S	N/S	1-1/4	1-5/16	1-9/16	1-9/16
N/R	108-33	108-31	N/R	N/R	N/R	N/R	N/R	N/S	1-3/16	N/R	1-1/2	N/R
N/R	108-33	108-31	N/R	N/R	N/R	N/R	N/R	N/S	1-3/16	N/R	1-1/2	N/R
Plain	108-33	108-31	108-30	108-30	108-13	N/S	N/S	N/S	1-3/32	1-3/32	1-1/2	1-1/2
Plain	108-33	108-31	108-30	108-30	108-13	N/S	N/S	N/S	1-3/32	1-3/32	1-1/2	1-1/2
Purple	108-33	108-31	108-30	108-30	108-13	N/S	N/S	N/S	1-1/4	1-5/16	1-9/16	1-9/16
Purple	108-33	108-31	108-30	108-30	108-13	N/S	N/S	N/S	1-1/4	1-5/16	1-9/16	1-9/16
Pink	108-33	108-31	108-30	108-30	108-13	N/S	N/S	N/S	1-3/8	1-7/16	1-11/16	1-11/16
N/R	108-33	108-31	N/R	N/R	N/R	N/R	N/R	N/S	1-3/16	N/R	1-1/2	N/R
N/S	108-33	108-31	N/R	N/R	N/R	N/R	N/R	N/S	1-3/16	N/R	1-1/2	N/R
Plain	108-33	108-31	108-30	108-30	108-13	N/S	N/S	N/S	1-3/32	1-3/32	1-1/2	1-1/2
Plain	108-33	108-31	108-30	108-30	108-13	N/S	N/S	N/S	1-3/32	1-3/32	1-1/2	1-1/2
Brown	108-33	108-31	108-30	108-30	108-13	N/S	N/S	N/S	1-1/4	1-5/16	1-9/16	1-9/16
Brown	108-33	108-31	108-30	108-30	108-13	N/S	N/S	N/S	1-1/4	1-5/16	1-9/16	1-9/16
Pink	108-33	108-31	108-30	108-30	108-13	N/S	N/S	N/S	1-1/4	1-5/16	1-9/16	1-9/16
Pink	(3)	(3)	(3)	(3)	(3)	N/R	N/R	N/R	1-1/4	1-1/4	1-11/16	1-11/16
Pink	(3)	(3)	(3)	(3)	(3)	N/R	N/R	N/R	1-1/4	1-1/4	1-11/16	1-11/16
.037	108-81	108-36	108-81	108-36	N/R	134-108	134-112	N/S	1-13/16	1-13/16	2	2
.037	108-81	108-36	108-81	108-36	N/R	134-108	134-112	N/S	1-14/16	1-14/16	2-1/8	2-1/8
N/R	108-39(5)	N/R	N/R	N/R	N/R	N/R	N/R	N/R	1-1/25	1-1/16	1-7/25	1-7/16
N/R	108-39(5)	N/R	N/R	N/R	N/R	N/R	N/R	N/R	1-1/25	1-1/16	1-7/25	1-7/16
N/R	108-39(5)	N/R	N/R	N/R	N/R	N/R	N/R	N/R	1-1/25	1-1/16	1-7/25	1-7/16
N/R	108-39(5)	N/R	N/R	N/R	N/R	N/R	N/R	N/R	1-1/25	1-1/16	1-7/25	1-7/16
Black	34-202	108-31	108-30	108-30	108-27	N/S	N/S	N/S	1-13/64	1-13/32	1-3/8	2
N/R	108-26	N/R	N/R	N/R	N/R	N/R	N/R	N/R	1-1/16	1-3/16	1-3/8	1-7/16
.028	108-33	108-29	N/R	N/R	N/R	N/R	N/R	N/R	1-3/8	1-3/8	1-11/16	1-11/16
Purple	108-56	108-55	108-30	108-30	108-13	N/S	N/S	N/S	1-1/4	1-5/16	1-9/16	1-9/16
Purple	108-56	108-55	108-30	108-30	108-13	N/S	N/S	N/S	1-1/4	1-5/16	1-9/16	1-9/16
Orange	108-56	108-55	108-30	108-30	108-13	N/S	N/S	N/S	1-1/4	1-5/16	1-9/16	1-9/16
Orange	108-56	108-55	108-30	108-30	108-13	N/S	N/S	N/S	1-1/4	1-5/16	1-9/16	1-9/16
.028	108-33	108-29	N/R	N/R	N/R	N/R	N/R	N/R	1-3/16	1-3/16	1-11/16	1-11/16
White	108-32	108-31	108-30	108-30	108-27	N/S	N/S	N/S	1-13/32	1-13/64	1-3/8	2
Orange	108-56	108-55	108-30	108-30	108-13	N/S	N/S	N/S	1-1/4	1-5/16	1-9/16	1-9/16
Brown	108-56	108-55	108-30	108-30	108-13	N/S	N/S	N/S	1-1/4	1-5/16	1-9/16	1-9/16
Brown	108-56	108-55	108-30	108-30	108-13	N/S	N/S	N/S	1-1/4	1-5/16	1-9/16	1-9/16
Brown	108-56	108-55	108-30	108-30	108-13	N/S	N/S	N/S	1-1/4	1-5/16	1-9/16	1-9/16
Brown	108-56	108-55	108-30	108-30	108-13	N/S	N/S	N/S	1-1/4	1-5/16	1-9/16	1-9/16
Black	108-32	108-31	108-30	108-30	108-27	N/S	N/S	N/S	1-13/32	1-13/64	1-3/8	2
Black	34-202	108-31	108-30	108-30	108-27	N/S	N/S	N/S	1-13/32	1-13/64	1-3/8	2
Plain	108-33	108-31	108-33	108-29	N/R	134-234	134-235	112-92	1-1/4	1-5/16	1-9/16	1-9/16
Black	108-32	108-31	108-30	108-30	108-27	N/S	N/S	N/S	1-13/32	1-13/64	1-3/8	2
Yellow	108-33	108-29	108-33	108-29	N/R	N/S	N/S	N/S	1-3/8	1-3/8	1-11/16	1-11/16
Purple	108-56	108-55	108-30	108-30	108-13	N/S	N/S	N/S	1-1/4	1-5/16	1-9/16	1-9/16
Brown	108-56	108-55	108-30	108-30	108-13	N/S	N/S	N/S	1-1/4	1-5/16	1-9/16	1-9/16
Pink	108-56	108-55	108-30	108-30	108-13	N/S	N/S	N/S	1-1/4	1-5/16	1-9/16	1-9/16
Pink	108-56	108-55	108-30	108-30	108-13	N/S	N/S	N/S	1-1/4	1-5/16	1-9/16	1-9/16
Black	108-32	108-35	108-30	108-30	108-27	N/S	N/S	N/S	1-13/32	1-13/64	1-3/8	2
Yellow	108-33	108-29	108-33	108-29	N/R	N/S	N/S	N/S	1-9/16	1-9/16	1-3/4	1-3/4
.035	108-33	108-29	108-33	108-29	N/R	134-108	134-112	N/S	1-11/16	1-11/16	2	2
.035	108-33	108-29	108-33	108-29	N/R	134-108	134-112	N/S	1-11/16	1-11/16	2	2

Super Tuning and Modifying Holley Carburetors

NUMERICAL LISTING AND COMPONENT PARTS

CARBURETOR PART NO.	CARB. MODEL NO.	CFM	RENEW KIT	TRICK KIT	PRIMARY & SECONDARY NEEDLE & SEAT	PRIMARY MAIN JET	SECONDARY MAIN JET OR PLATE	PRIMARY METERING BLOCK	SECONDARY METERING BLOCK	PRIMARY POWER VALVE	PRIMARY DISCHARGE NOZZLE SIZE
R80262	4160	650	703-28	N/A	6-516	122-74	N/S	N/S	N/R	125-65	.040
R80263	2300	300	703-30	N/A	6-516	122-60	N/R	N/S	N/R	125-50	.028
R80264	4160	450	703-28	N/A	6-516	122-59	N/S	N/S	N/S	125-85	.021
R80265	4160	600	703-29	N/A	6-516	122-65	N/S	N/S	N/R	125-25	.025
R80309	4150	715	703-45	N/A	6-504	122-72	(1)	N/S	N/S	125-25	.031
R80310	4175	650	703-34	N/A	6-511	122-61	N/S	N/S	N/R	125-50	.040
R80310-1	4175	650	703-34	N/A	6-511	122-61	N/S	N/S	N/R	125-50	.040
R80310-2	4175	650	703-34	N/A	6-511	122-61	N/S	N/S	N/R	125-50	.040
R80311	4150	850	703-35	N/A	6-504	122-84	122-88	N/S	N/S	125-65(22)	.040
R80311-1	4150	850	703-35	N/A	6-504	122-84	122-88	N/S	N/S	125-65(22)	.040
R80311-2	4150	850	703-35	N/A	6-504	122-84	122-88	N/S	N/S	125-65(22)	.040
R80312	2300	350	703-36	N/A	6-511	122-70	N/R	N/S	N/R	125-25	.028
R80312-1	2300	350	703-36	N/A	6-511	122-70	N/R	N/S	N/R	125-25	.028
R80313	2300	350	703-41	N/A	6-506	122-62	N/R	N/S	N/R	125-25	.031
R80313-1	2300	350	703-41	N/A	6-506	122-62	N/R	N/S	N/R	125-25	.031
R80315	4160	600	703-29	N/A	6-511	122-67	N/S	N/S	N/R	125-25	.025
R80315-1	4160	600	703-29	N/A	6-511	122-67	N/S	N/S	N/R	125-25	.025
R80316	2300	500	703-41	N/A	6-506	122-75	N/R	N/S	N/R	125-25	.028
R80316-1	2300	500	703-41	N/A	6-506	122-75	N/R	N/S	N/R	125-25	.028
R80318-1	4160	600	703-33	N/A	6-516	122-74	N/S	N/S	N/S	125-65	.040
R80319-1	4160	600	703-47	N/A	6-511	122-65	N/S	N/S	N/S	125-25	.025
R80320-1	2300	350	703-30	N/A	6-516	122-61	N/R	N/S	N/R	125-50	.028
R80321	2300	350	703-41	N/A	6-506	122-63	N/R	N/S	N/R	125-35	.031
R80321-1	2300	350	703-41	N/A	6-506	122-63	N/R	N/S	N/R	125-35	.031
R80328	4175	650	703-40	N/A	6-511	122-62	N/S	N/S	N/R	125-50	.040
R80328-1	4175	650	703-40	N/A	6-511	122-62	N/S	N/S	N/R	125-50	.040
R80328-2	4175	650	703-40	N/A	6-511	122-62	N/S	N/S	N/R	125-50	.040
R80330	4150	850	703-35	N/A	6-504	122-88	122-94	N/S	N/S	125-65	.040
R80330-1	4150	850	703-35	N/A	6-504	122-88	122-94	N/S	N/S	125-65	.040
R80340	4500	1050	703-44	N/A	6-504	122-84	122-84	N/S	N/S	125-65(15)	.035
R80340-1	4500	1050	703-44	37-920	6-518	122-88	122-88	134-263	134-263	N/R	.035
R80341	4160	390	703-17	N/A	6-506	122-54	134-32	N/S	N/S	125-65	.059
R80364	4160	450	703-28	N/A	6-511	122-59	N/S	N/S	N/S	125-85	.021
R80378	4150	750	703-48	N/A	6-511	122-59	122-72	N/S	N/S	125-25(30)	.035
R80378-1	4150	750	703-48	N/A	6-511	122-56	122-73	N/S	N/S	125-25(30)	.031
R80382	2300	350	703-49	N/A	6-511	122-71	N/R	N/S	N/R	125-35	.031
R80382-1	2300	350	703-49	N/A	6-511	122-71	N/R	N/S	N/R	125-35	.031
R80382-2	2300	350	703-49	N/A	6-511	122-71	N/R	N/S	N/R	125-35	.037
R80383	4160	650	703-47	N/A	6-511	122-68	N/S	N/S	N/R	125-25	.035
R80383-1	4160	650	703-47	N/A	6-511	122-68	N/S	N/S	N/R	125-25	.035
R80385	2300	350	703-41	N/A	6-506	122-75	N/R	N/S	N/R	125-25	.028
R80386	2300	350	703-49	N/A	6-511	122-61	N/R	N/S	N/R	125-35	.028
R80386-1	2300	350	703-49	N/A	6-511	122-61	N/R	N/S	N/R	125-35	.028
R80390	4175	650	703-50	N/A	6-511	122-61	134-22	N/S	N/R	125-25	.040
R80391	4160	700	703-34	N/A	6-511	122-69	N/S	N/S	N/R	125-25	.035
R80402	2300	450	703-36	N/A	6-511	122-75	N/R	N/S	N/R	125-25	.028
R80402-1	2300	450	703-36	N/A	6-511	122-75	N/R	N/S	N/R	125-45	.028
R80403	4160	600	703-29	N/A	6-511	122-64	N/S	N/S	N/R	125-25	.025
R80403-1	4160	600	703-29	N/A	6-511	122-64	N/S	N/S	N/R	125-25	.032
R80408	4150	715	703-45	N/A	6-504	122-73	(1)	N/S	N/R	125-25	.031
R80427	4150	700	3-485	N/A	6-504	122-74	122-84	N/S	N/S	125-45	.037
R80431	4160	550	37-119	37-912	6-506	122-60	134-9	N/S	N/S	125-65	.025
R80432	4160	550	37-119	37-912	6-506	122-60	134-9	N/S	N/S	125-65	.025
R80434	4160	750	703-55	N/A	6-516	122-69	N/S	N/S	N/R	125-45	.035
R80436	4150	850	37-1184	37-915	6-504	122-80	122-80	134-231	134-246	125-65(22)	.040
R80443	4150	850	703-58	N/A	6-504	122-88	122-96	N/S	N/S	125-65(15)	.031
R80444	4150	850	703-35	N/A	6-504	122-88	122-94	N/S	N/S	125-65(22)	.040
R80450	4160	600	37-1415	37-912	6-506	122-622	134-39	N/S	N/S	125-208	.031
R80451	4160	600	37-1415	37-912	6-506	122-622	134-39	N/S	N/S	125-208	.031
R80452	4160	600	37-1415	37-912	6-506	122-652	134-39	N/S	N/S	125-208	.031
R80453	4160	600	37-1415	37-912	6-506	122-632	134-39	N/S	N/S	125-208	.031
R80454	4160	600	37-1415	37-912	6-506	122-622	134-39	N/S	N/S	125-208	.031
R80456-1	4160	600	703-61	N/A	6-511	122-65	N/S	N/S	N/S	125-25	.025

NUMERICAL LISTING AND COMPONENT PARTS

SECONDARY NOZZLE SIZE OR SPRING COLOR	PRIMARY BOWL GASKET†	PRIMARY METERING BLOCK GASKET†	SECONDARY BOWL GASKET†	SECONDARY METERING BLOCK GASKET†	SECONDARY METERING PLATE GASKET†	PRIMARY FUEL BOWL	SECONDARY FUEL BOWL	THROTTLE BODY & SHAFT ASSEMBLY	VENTURI DIAMETER PRIMARY	VENTURI DIAMETER SECONDARY	THROTTLE BORE DIAMETER PRIMARY	THROTTLE BORE DIAMETER SECONDARY
Pink	108-33	108-31	108-30	108-30	108-13	N/S	N/S	N/S	1-1/4	1-5/16	1-9/16	1-9/16
N/R	108-33	108-31	N/R	N/R	N/R	N/S	N/R	N/R	1-3/16	N/R	1-1/2	N/R
Plain	108-33	108-31	108-30	108-30	108-13	N/S	N/S	N/S	1-3/32	1-3/32	1-1/2	1-1/2
Purple	108-33	108-31	108-30	108-30	108-13	N/S	N/S	N/S	1-1/4	1-5/16	1-9/16	1-9/16
Yellow	108-33	108-29	108-33	108-29	N/R	N/S	N/S	N/S	1-3/8	1-3/8	1-11/16	1-11/16
Red	108-32	108-31	108-30	108-30	108-27	N/S	N/S	N/S	1-13/64	1-13/64	1-3/8	2
Red	108-32	108-31	108-30	108-30	108-27	N/S	N/S	N/S	1-13/64	1-13/64	1-3/8	2
Red	108-32	108-31	108-30	108-30	108-27	N/S	N/S	N/S	1-13/64	1-13/64	1-3/8	2
Yellow	108-33	108-29	108-33	108-29	N/R	N/S	N/S	N/S	1-9/16	1-9/16	1-3/4	1-3/4
Yellow	108-33	108-29	108-33	108-29	N/R	N/S	N/S	N/S	1-9/16	1-9/16	1-3/4	1-3/4
Yellow	108-33	108-29	108-33	108-29	N/R	N/S	N/S	N/S	1-9/16	1-9/16	1-3/4	1-3/4
N/R	108-33	108-29	N/R	N/R	N/R	N/S	N/R	N/R	1-3/16	N/R	1-1/2	N/R
N/R	108-33	108-29	N/R	N/R	N/R	N/S	N/R	N/R	1-3/16	N/R	1-1/2	N/R
N/R	108-33	108-29	N/R	N/R	N/R	N/S	N/R	N/R	1-3/16	N/R	1-1/2	N/R
Yellow	108-33	108-31	108-30	108-30	108-27	N/S	N/S	N/S	1-1/4	1-5/16	1-9/16	1-9/16
Yellow	108-33	108-31	108-30	108-30	108-27	N/S	N/S	N/S	1-1/4	1-5/16	1-9/16	1-9/16
N/R	108-33	108-29	N/R	N/R	N/R	N/R	N/R	N/R	1-3/8	N/R	1 11/16	N/R
N/R	108-33	108-33	N/R	N/R	N/R	N/R	N/R	N/R	1-3/8	N/R	1-11/16	N/R
Pink	108-33	108-31	108-30	108-30	108-13	N/S	N/S	N/S	1 1/4	1-5/16	1-9/16	1-9/16
Brown	108-33	108-31	108-30	108-30	108-13	N/S	N/S	N/S	1-1/4	1-5/16	1-9/16	1-9/16
N/R	108-33	108-31	N/R	N/R	N/R	N/R	N/R	N/R	1-3/16	N/R	1-1/2	N/R
N/R	108-33	108-31	N/R	N/R	N/R	N/R	N/R	N/R	1-3/16	N/R	1-1/2	N/R
N/R	108-33	108-31	N/R	N/R	N/R	N/R	N/R	N/R	1-3/16	N/R	1-1/2	N/R
Red	108-32	108-31	108-30	108-30	108-27	N/S	N/S	N/S	1-13/64	1-13/64	1-3/8	2
Red	108-32	108-31	108-30	108-30	108-27	N/S	N/S	N/S	1-13/64	1-13/64	1-3/8	2
Red	108-32	108-31	108-30	108-30	108-27	N/S	N/S	N/S	1-13/64	1-13/64	1-3/8	2
Pink	108-33	108-29	108-33	108-29	N/R	N/S	N/S	N/S	1-9/16	1-9/16	1-3/4	1-3/4
Pink	108-33	108-29	108-33	108-29	N/R	N/S	N/S	N/S	1-9/16	1-9/16	1-3/4	1-3/4
N/R	108-33	108-29	108-33	108-29	N/R	N/S	N/S	N/R	1-11/16	1-11/16	2	2
.037	108-81	108-36	108-81	108-36	N/R	134-108	134-112	N/S	1-11/16	1-11/16	2	2
Yellow	108-33	108-31	108-30	108-30	108-27	N/S	N/S	N/S	1-1/16	1-1/16	1-7/16	1-7/16
Plain	108-33	108-31	108-30	108-30	108-27	N/S	N/S	N/S	1-3/32	1-3/32	1-1/2	1-1/2
Pink	108-33	108-29	108-33	108-29	N/R	N/S	N/S	N/S	1-3/8	1-7/16	1-11/16	1-11/16
Pink	108-33	108-29	108-33	108-29	N/R	N/S	N/S	N/S	1-3/8	1-7/16	1-11/16	1-11/16
N/R	108-33	108-29	N/R	N/R	N/R	N/R	N/R	N/R	1-3/16	N/R	1-1/2	N/R
N/R	108-33	108-29	N/R	N/R	N/R	N/R	N/R	N/R	1-3/16	N/R	1-1/2	N/R
N/R	108-33	108-29	N/R	N/R	N/R	N/R	N/R	N/R	1-3/16	N/R	1-1/2	N/R
Brown	108-33	108-31	108-30	108-30	108-13	N/S	N/S	N/S	1-1/4	1-5/16	1-9/16	1-9/16
Brown	108-33	108-31	108-30	108-30	108-13	N/S	N/S	N/S	1-1/4	1-5/16	1-9/16	1-9/16
N/R	108-33	108-29	N/R	N/R	N/R	N/R	N/R	N/R	1-3/16	N/R	1-1/2	N/R
N/R	108-33	108-31	N/R	N/R	N/R	N/R	N/R	N/R	1-3/16	N/R	1-1/2	N/R
Red	108-32	108-31	108-30	108-30	108-27	N/S	N/S	N/S	1-13/64	1-13/32	1-3/8	2
Yellow	108-33	108-31	108-30	108-30	108-27	N/S	N/S	N/S	1-9/16	N/R	1-3/4	N/R
N/R	108-33	108-29	N/R	N/R	N/R	N/S	N/R	N/R	1-9/16	N/R	1-3/4	N/R
N/R	108-33	108-29	N/R	N/R	N/R	N/S	N/R	N/R	1-9/16	N/R	1-3/4	N/R
Red	108-33	108-31	108-30	108-30	108-27	N/S	N/S	N/S	1-1/4	1-5/16	1-9/16	1-9/16
Red	108-33	108-31	108-30	108-30	108-27	N/S	N/S	N/S	1-1/4	1-5/16	1-9/16	1-9/16
Yellow	108-33	108-29	108-33	108-29	N/R	N/S	N/S	N/S	1-5/16	1-3/8	1-11/16	1-11/16
.037	108-33	108-29	108-33	108-29	N/R	N/S	N/S	N/S	1-5/16	1-3/8	1-11/16	1-11/16
Plain	108-33	108-29	108-30	108-30	108-27	N/S	N/S	N/S	1-3/16	1-1/4	1-1/2	1-1/2
Plain	108-33	108-29	108-30	108-30	108-27	N/S	N/S	N/S	1-3/16	1-1/4	1-1/2	1-1/2
Orange/Red	108-33	108-31	108-30	108-30	108-13	N/S	N/S	N/S	1-3/8	1-3/8	1-11/16	1-11/16
Pink	108-33	108-29	108-33	108-29	N/R	134-103	134-102	112-95	1-9/16	1-9/16	1-3/4	1-3/4
.031	108-33	108-29	108-33	108-29	N/R	N/S	N/S	N/S	1-9/16	1-9/16	1-3/4	1-3/4
Yellow	108-33	108-29	108-33	108-29	N/R	N/S	N/S	N/S	1-9/16	1-9/16	1-3/4	1-3/4
Black	108-33	108-29	108-30	108-30	N/R	N/S	N/S	N/S	1-1/4	1-5/16	1-9/16	1-9/16
Black	108-33	108-29	108-30	108-30	N/R	N/S	N/S	N/S	1-1/4	1-5/16	1-9/16	1-9/16
Black	108-33	108-29	108-30	108-30	N/R	N/S	N/S	N/S	1-1/4	1-5/16	1-9/16	1-9/16
Black	108-33	108-29	108-30	108-30	N/R	N/S	N/S	N/S	1-1/4	1-5/16	1-9/16	1-9/16
Black	108-33	108-29	108-30	108-30	N/R	N/S	N/S	N/S	1-1/4	1-5/16	1-9/16	1-9/16
Brown	108-33	108-31	108-30	108-30	108-13	N/S	N/S	N/S	1-1/4	1-5/16	1-9/16	1-9/16

Super Tuning and Modifying Holley Carburetors

NUMERICAL LISTING AND COMPONENT PARTS

CARBURETOR PART NO.	CARB. MODEL NO.	CFM	RENEW KIT	TRICK KIT	PRIMARY & SECONDARY NEEDLE & SEAT	PRIMARY MAIN JET	SECONDARY MAIN JET OR PLATE	PRIMARY METERING BLOCK	SECONDARY METERING BLOCK	PRIMARY POWER VALVE	PRIMARY DISCHARGE NOZZLE SIZE
R80457	4160	600	37-119	37-912	6-506	122-69	134-39	134-128	134-39	125-65	.031
R80457-1	4160	600	37-119	37-912	6-506	122-64	134-39	134-128	134-39	125-65	.031
R80460	4160	600	37-1415	37-912	6-506	122-622	134-39	N/S	N/S	125-208	.031
R80466	4150	800	N/A	N/A	6-504	122-72	122-87	N/S	N/S	125-45	.031
R80473	4160	600	N/A	N/A	6-511	122-64	N/S	N/S	N/S	125-25	.025
R80473-1	4160	600	N/A	N/A	6-511	122-64	N/S	N/S	N/S	125-25	.037
R80487	4160	600	703-66	N/A	6-511	122-68	N/S	N/S	N/S	125-45	.037
R80491	4175	650	37-741	N/A	6-511	122-632	134-21	134-201	134-21	1025-475-13	.037
R80492	4160	600	703-29	N/A	6-511	122-68	N/S	N/S	N/S	125-45	.037
R80496	4150	950	37-1532	N/A	6-518	122-78	122-78	134-252	134-252	125-165(both)	.031
R80497	4150	950	37-1532	N/A	6-518	122-78	122-78	134-252	134-252	125-165(both)	.031
R80498	4150	950	37-1533	N/A	6-519	122-144	122-144	134-253	134-253	125-155(both)	.055
R80502	2300	500	703-62	N/A	6-504	122-71	N/R	N/S	N/R	125-35	.047
R80507	4150	390	37-1530	N/A	6-504	122-65	122-65	134-254	134-254	125-35(22)	.025
R80508	4160	750	37-754	37-921	6-504	122-72	134-21	134-131	134-21	125-65	.025
R80509	4150	830	37-1531	N/A	6-504	122-86	122-86	134-255	134-255	125-65(15)	.028
R80511	4150	830	37-1532	N/A	6-518	122-84	122-84	134-256	134-256	125-65(15)	.028
R80512	4150	1000	37-1532	N/A	6-518	122-84	122-84	134-257	134-257	125-65(15)	.031
R80513	4150	1000	37-1532	N/A	6-518	122-84	122-84	134-257	134-257	125-65(15)	.031
R80514	4150	1000	37-1532	N/A	6-518	122-84	122-88	134-258	134-258	125-65(15)	.031
R80519	4150	1000	37-1532	N/A	6-518	122-84	122-88	134-258	134-258	125-65(15)	.031
R80528	4150	750	37-1531	N/A	6-504	122-72	122-84	134-261	134-261	125-65	.031
R80529	4150	750	37-1531	N/A	6-504	122-72	122-84	134-261	134-261	125-65	.031
R80532	4500	1250	37-487	37-920	6-518	122-97	122-97	134-265	134-265	N/R	.035
R80533	4500	1250	37-487	37-920	6-518	122-97	122-97	134-265	134-265	N/R	.035
R80535	4150	750	37-1534	N/A	6-519	122-132	122-132	134-262	134-262	125-55	.045
R80537	4150	750	37-1535	N/A	6-504	122-73	122-81	N/S	134-250	125-65	.028
R80776	4150	600	34-485	37-910	6-504	122-66	122-73	134-145	134-143	125-65	.028
R80777	4150	650	37-485	37-910	6-504	122-67	122-73	134-150	134-147	125-65	.028
R80778	4150	700	37-485	37-910	6-504	122-69	122-78	134-225	134-226	125-65	.028
R80779	4150	750	37-485	37-910	6-504	122-70	122-80	134-227	134-250	125-65	.028
R80780	4150	800	37-458	37-910	6-504	122-71	122-85	134-229	134-230	125-65	.031
R80781	4150	850	37-485	37-916	6-504	122-80	122-78	134-249	134-212	125-65 (15)	.031
R81850	4160	600	37-119	37-912	6-506	122-66	134-9	134-128	134-9	125-65	.025
R82010	2010	350	37-1467	37-931	6-504	122-58	N/A	N/R	N/R	125-65	.035
R82011	2010	500	37-1468	37-932	6-504	122-80	N/A	N/R	N/R	125-65	.035
R82012	2010	560	37-1468	37-932	6-504	122-80	N/A	N/R	N/R	125-65	.035
R82020	2010	350	703-51	N/A	6-504	122-60	N/A	N/R	N/R	125-65	.035
R82021	2010	500	703-51	N/A	6-504	122-80	N/A	N/R	N/R	125-65	.035
R82028	2010	500	703-51	N/A	6-504	122-80	N/A	N/R	N/R	125-65	.035
R82028-1	2010	500	703-51	N/A	6-504	122-80	N/A	N/R	N/R	125-65	.035
R82029	2010	500	703-51	N/A	6-504	122-81	N/A	N/R	N/R	125-65	.035
R83310	4160	750	37-754	37-921	6-504	122-72	134-21	134-131	134-21	125-65	.025
R83310-1	4160	750	37-754	37-921	6-504	122-72	134-21	134-131	134-21	125-65	.025
R83311	4160	750	37-754	37-921	6-504	122-72	134-21	134-131	134-21	125-65	.025
R83312	4160	750	37-754	37-921	6-504	122-72	134-21	134-131	134-21	125-65	.025
R84010	4010	600	37-1445	37-927	6-504	122-67	122-75	N/R	N/R	125-65	.026
R84010-1	4010	600	37-1445	37-927	6-504	122-67	122-75	N/R	N/R	125-65	.035
R84010-2	4010	600	37-1445	37-927	6-504	122-67	122-75	N/R	N/R	125-65	.035
R84010-3	4010	600	37-1445	37-927	6-504	122-63	122-75	N/R	N/R	125-65	.035
R84011	4010	750	37-1445	37-927	6-504	122-75	122-75	N/R	N/R	125-65(15)	.026
R84011-1	4010	750	37-1445	37-927	6-504	122-75	122-75	N/R	N/R	125-65(15)	.035
R84011-2	4010	750	37-1445	37-927	6-504	122-75	122-75	N/R	N/R	125-65	.035
R84011-3	4010	750	37-1445	37-927	6-504	122-73	122-75	N/R	N/R	125-65	.031
R84012	4010	600	37-1446	37-928	6-504	122-67	122-77	N/R	N/R	125-65	.026
R84012-1	4010	600	37-1446	37-928	6-504	122-67	122-77	N/R	N/R	125-65	.035
R84012-2	4010	600	37-1446	37-928	6-504	122-67	122-77	N/R	N/R	125-65	.035
R84012-3	4010	600	37-1446	37-928	6-504	122-67	122-77	N/R	N/R	125-65	.031
R84013	4010	750	37-1446	37-928	6-504	122-79	122-79	N/R	N/R	125-65(15)	.026
R84013-1	4010	750	37-1446	37-928	6-504	122-79	122-79	N/R	N/R	125-65(15)	.035
R84013-2	4010	750	37-1446	37-928	6-504	122-79	122-79	N/R	N/R	125-65(15)	.035
R84013-3	4010	750	37-1446	37-928	6-504	122-75	122-79	N/R	N/R	125-65	.031
R84014	4011	650	37-1447	37-929	6-504	122-60	122-66	N/R	N/R	125-65(15)	.026

NUMERICAL LISTING AND COMPONENT PARTS

SECONDARY NOZZLE SIZE OR SPRING COLOR	PRIMARY BOWL GASKET†	PRIMARY METERING BLOCK GASKET†	SECONDARY BOWL GASKET†	SECONDARY METERING BLOCK GASKET†	SECONDARY METERING PLATE GASKET†	PRIMARY FUEL BOWL	SECONDARY FUEL BOWL	THROTTLE BODY & SHAFT ASSEMBLY	VENTURI DIAMETER PRIMARY	VENTURI DIAMETER SECONDARY	THROTTLE BORE DIAMETER PRIMARY	THROTTLE BORE DIAMETER SECONDARY
Purple	N/A	N/A	N/A	N/A	N/A	N/R	N/R	N/R	1-1/4	1-1/2	1-11/16	1-11/16
Purple	N/A	N/A	N/A	N/A	N/A	N/R	N/R	N/R	1-1/2	1-1/2	1-11/16	1-11/16
Purple	N/A	N/A	N/A	N/A	N/A	N/R	N/R	N/R	1-1/2	1-1/2	1-11/16	1-11/16
Purple	N/A	N/A	N/A	N/A	N/A	N/R	N/R	N/R	1-1/2	1-1/2	1-11/16	1-11/16
Purple	N/A	N/A	N/A	N/A	N/A	N/R	N/R	N/R	1-1/2	1-1/2	1-11/16	1-11/16
.026	N/A	N/A	N/A	N/A	N/A	N/R	N/R	N/R	1-1/4	1-1/4	1-11/16	1-11/16
.026	N/A	N/A	N/A	N/A	N/A	N/R	N/R	N/R	1-1/4	1-1/4	1-11/16	1-11/16
.026	N/A	N/A	N/A	N/A	N/A	N/R	N/R	N/R	1-1/4	1-1/4	1-11/16	1-11/16
.026	(3)	(3)	(3)	(3)	(3)	N/R	N/R	N/R	1-1/4	1-1/4	1-11/16	1-11/16
.026	(3)	(3)	(3)	(3)	(3)	N/R	N/R	N/R	1-1/2	1-1/2	1-11/16	1-11/16
.026	(3)	(3)	(3)	(3)	(3)	N/R	N/R	N/R	1-1/2	1-1/2	1-11/16	1-11/16
.026	(3)	(3)	(3)	(3)	(3)	N/R	N/R	N/R	1-1/2	1-1/2	1-11/16	1-11/16
.026	(3)	(3)	(3)	(3)	(3)	N/R	N/R	N/R	1-1/2	1-1/2	1-11/16	1-11/16
Plain	(4)	(4)	(4)	(4)	(4)	N/R	N/R	N/R	1-5/32	1-3/8	1-3/8	2
Plain	(4)	(4)	(4)	(4)	(4)	N/R	N/R	N/R	1-5/32	1-3/8	1-3/8	2
Plain	(4)	(4)	(4)	(4)	(4)	N/R	N/R	N/R	1-5/32	1-3/8	1-3/8	2
Plain	(4)	(4)	(4)	(4)	(4)	N/R	N/R	N/R	1-5/32	1-3/8	1-3/8	2
Yellow	(4)	(4)	(4)	(4)	(4)	N/R	N/R	N/R	1-5/32	1-23/32	1-3/8	2
Yellow	(4)	(4)	(4)	(4)	(4)	N/R	N/R	N/R	1-5/32	1-23/32	1-3/8	2
Yellow	(4)	(4)	(4)	(4)	(4)	N/R	N/R	N/R	1-5/32	1-23/32	1-3/8	2
Plain	(4)	(4)	(4)	(4)	(4)	N/R	N/R	N/R	1-5/32	1-23/32	1-3/8	2
.026	(4)	(4)	(4)	(4)	(4)	N/R	N/R	N/R	1-5/32	1-3/8	1-3/8	2
.026	(4)	(4)	(4)	(4)	(4)	N/R	N/R	N/R	1-5/32	1-3/8	1-3/8	2
.026	(4)	(4)	(4)	(4)	(4)	N/R	N/R	N/R	1-5/32	1-3/8	1-3/8	2
.026	(4)	(4)	(4)	(4)	(4)	N/R	N/R	N/R	1-5/32	1-23/32	1-3/8	2
.026	(4)	(4)	(4)	(4)	(4)	N/R	N/R	N/R	1-5/32	1-23/32	1-3/8	2
.026	(4)	(4)	(4)	(4)	(4)	N/R	N/R	N/R	1-5/32	1-23/32	1-3/8	2
Yellow	(3)	(3)	(3)	(3)	(3)	N/R	N/R	N/R	1-1/2	1-1/2	1-11/16	1-11/16
Yellow	(3)	(3)	(3)	(3)	(3)	N/R	N/R	N/R	1-1/2	1-1/2	1-11/16	1-11/16
Yellow	(3)	(3)	(3)	(3)	(3)	N/R	N/R	N/R	1-1/2	1-1/2	1-11/16	1-11/16
Purple	(3)	(3)	(3)	(3)	(3)	N/R	N/R	N/R	1-1/4	1-1/4	1-11/16	1-11/16
Purple	(3)	(3)	(3)	(3)	(3)	N/R	N/R	N/R	1-1/4	1-1/4	1-11/16	1-11/16
Purple	(3)	(3)	(3)	(3)	(3)	N/R	N/R	N/R	1-1/4	1-1/4	1-11/16	1-11/16
Plain	(4)	(4)	(4)	(4)	(4)	N/R	N/R	N/R	1-5/32	1-3/8	1-3/8	2
Plain	(4)	(4)	(4)	(4)	(4)	N/R	N/R	N/R	1-5/32	1-3/8	1-3/8	2
Plain	(4)	(4)	(4)	(4)	(4)	N/R	N/R	N/R	1-5/32	1-3/8	1-3/8	2
Plain	(4)	(4)	(4)	(4)	(4)	N/R	N/R	N/R	1-5/32	1-3/8	1-3/8	2
Yellow	(4)	(4)	(4)	(4)	(4)	N/R	N/R	N/R	1-5/32	1-23/32	1-3/8	2
Yellow	(4)	(4)	(4)	(4)	(4)	N/R	N/R	N/R	1-5/32	1-23/32	1-3/8	2
Yellow	(3)	(3)	(3)	(3)	(3)	N/R	N/R	N/R	1-1/4	1-1/4	1-11/16	1-11/16
Yellow	(3)	(3)	(3)	(3)	(3)	N/R	N/R	N/R	1-1/4	1-1/4	1-11/16	1-11/16
Yellow	(3)	(3)	(3)	(3)	(3)	N/R	N/R	N/R	1-1/4	1-1/4	1-11/16	1-11/16
Yellow	(4)	(4)	(4)	(4)	(4)	N/R	N/R	N/R	1-5/32	1-3/8	1-3/8	2
Yellow	(4)	(4)	(4)	(4)	(4)	N/R	N/R	N/R	1-5/32	1-3/8	1-3/8	2
Yellow	(4)	(4)	(4)	(4)	(4)	N/R	N/R	N/R	1-5/32	1-3/8	1-3/8	2
Yellow	**(4)**	**(4)**	**(4)**	**(4)**	**(4)**	**N/R**	**N/R**	**N/R**	**1-5/32**	**1-3/8**	**1-3/8**	**2**
Red	(3)	(3)	(3)	(3)	(3)	N/R	N/R	N/R	1-1/2	1-1/2	1-11/16	1-11/16
Purple	(3)	(3)	(3)	(3)	(3)	N/R	N/R	N/R	1-1/4	1-1/4	1-11/16	1-11/16
Purple	(3)	(3)	(3)	(3)	(3)	N/R	N/R	N/R	1-1/4	1-1/4	1-11/16	1-11/16
Purple	(3)	(3)	(3)	(3)	(3)	N/R	N/R	N/R	1-1/4	1-1/4	1-11/16	1-11/16
Yellow	(4)	(4)	(4)	(4)	(4)	N/R	N/R	N/R	1-5/32	1-3/8	1-3/8	2
.035	(3)	(3)	(3)	(3)	(3)	N/R	N/R	N/R	1-1/4	1-1/4	1-11/16	1-11/16
.035	(3)	(3)	(3)	(3)	(3)	N/R	N/R	N/R	1-1/2	1-1/2	1-11/16	1-11/16
.026	(4)	(4)	(4)	(4)	(4)	N/R	N/R	N/R	1-5/32	1-3/8	1-3/8	2
.026	(4)	(4)	(4)	(4)	(4)	N/R	N/R	N/R	1-5/32	1-3/8	1-3/8	2
.026	(4)	(4)	(4)	(4)	(4)	N/R	N/R	N/R	1-5/32	1-28/32	1-3/8	2
.026	(4)	(4)	(4)	(4)	(4)	N/R	N/R	N/R	1-5/32	1-28/32	1-3/8	2
Yellow	(4)	(4)	(4)	(4)	(4)	N/R	N/R	N/R	1-5/32	1-3/8	1-3/8	2
Red	(3)	(3)	(3)	(3)	(3)	N/R	N/R	N/R	1-1/2	1-1/2	1-11/16	1-11/1

NUMERICAL LISTING AND COMPONENT PARTS

CARBURETOR PART NO.	CARB. MODEL NO.	CFM	RENEW KIT	TRICK KIT	PRIMARY & SECONDARY NEEDLE & SEAT	PRIMARY MAIN JET	SECONDARY MAIN JET OR PLATE	PRIMARY METERING BLOCK	SECONDARY METERING BLOCK	PRIMARY POWER VALVE	PRIMARY DISCHARGE NOZZLE SIZE
R84014-1	4011	650	37-1447	37-929	6-504	122-60	122-66	N/R	N/R	125-65(15)	.026
R84014-2	4011	650	37-1447	37-929	6-504	122-60	122-64	N/R	N/R	125-65(15)	.026
R84014-3	4011	650	37-1447	37-929	6-504	122-60	122-64	N/R	N/R	125-65(15)	.026
R84015	4011	800	37-1447	37-929	6-504	122-64	122-90	N/R	N/R	125-65(15)	.026
R84015-1	4011	800	37-1447	37-929	6-504	122-64	122-90	N/R	N/R	125-65(15)	.026
R84015-2	4011	800	37-1447	37-929	6-504	122-60	122-90	N/R	N/R	125-65(15)	.026
R84015-3	4011	800	37-1447	37-929	6-504	122-60	122-90	N/R	N/R	125-65(15)	.026
R84016	4011	650	37-1448	37-930	6-504	122-64	122-64	N/R	N/R	125-65(15)	.026
R84016-1	4011	650	37-1448	37-930	6-504	122-64	122-64	N/R	N/R	125-65(15)	.026
R84016-2	4011	650	37-1448	37-930	6-504	122-60	122-64	N/R	N/R	125-65(15)	.026
R84016-3	4011	650	37-1448	37-930	6-504	122-60	122-64	N/R	N/R	125-65(15)	.026
R84017	4011	800	37-1448	37-930	6-504	122-64	122-90	N/R	N/R	125-65(15)	.026
R84017-1	4011	800	37-1448	37-930	6-504	122-64	122-90	N/R	N/R	125-65(15)	.026
R84017-2	4011	800	37-1448	37-930	6-504	122-60	122-90	N/R	N/R	125-65(15)	.026
R84017-3	4011	800	37-1448	37-930	6-504	122-60	122-90	N/R	N/R	125-65(15)	.026
R84018	4010	750	37-1445	N/A	6-504	122-86	122-86	N/R	N/R	125-65(15)	.026
R84018-1	4010	750	37-1445	N/A	6-504	122-86	122-86	N/R	N/R	125-65(15)	.026
R84018-2	4010	750	37-1445	N/A	6-504	122-86	122-86	N/R	N/R	125-65(15)	.035
R84020	4010	600	37-1445	37-927	6-504	122-67	122-75	N/R	N/R	125-65	.026
R84020-1	4010	600	37-1445	37-927	6-504	122-67	122-75	N/R	N/R	125-65	.035
R84020-2	4010	600	37-1445	37-927	6-504	122-67	122-75	N/R	N/R	125-65	.035
R84020-3	4010	600	37-1445	37-927	6-504	122-63	122-75	N/R	N/R	125-65	.035
R84021	4011	650	37-1447	37-929	6-504	122-60	122-64	N/R	N/R	125-65(15)	.026
R84021-1	4011	650	37-1447	37-929	6-504	122-60	122-64	N/R	N/R	125-65(15)	.026
R84021-2	4011	650	37-1447	37-929	6-504	122-60	122-64	N/R	N/R	125-65(15)	.026
R84021-3	4011	650	37-1447	37-929	6-504	122-60	122-64	N/R	N/R	125-65(15)	.026
R84022	4011	800	37-1447	N/A	6-504	122-64	122-95	N/R	N/R	125-65	.026
R84022-1	4011	800	37-1447	N/A	6-504	122-64	122-95	N/R	N/R	125-65(15)	.026
R84023	4010	600	37-1445	N/A	6-504	122-67	122-75	N/R	N/R	125-65	.026
R84023-1	4010	600	37-1445	N/A	6-504	122-67	122-75	N/R	N/R	125-65	.026
R84023-2	4010	600	37-1445	N/A	6-504	122-67	122-75	N/R	N/R	125-65	.035
R84024	4011	650	37-1447	N/A	6-504	122-60	122-64	N/R	N/R	125-65(15)	.026
R84024-1	4011	650	37-1447	N/A	6-504	122-64	122-68	N/R	N/R	125-65(15)	.026
R84026	4011	650	37-1447	N/A	6-504	122-64	122-68	N/R	N/R	125-65(15)	.026
R84026-1	4011	650	37-1447	N/A	6-504	122-64	122-68	N/R	N/R	125-65(15)	.026
R84026-2	4011	650	37-1447	N/A	6-504	122-64	122-68	N/R	N/R	125-65(15)	.026
R84028	4010	750	37-1445	N/A	6-504	122-86	122-90	N/R	N/R	125-65(15)	.035
R84035	4010	600	37-1445	37-927	6-504	122-67	122-75	N/R	N/R	125-65	.035
R84035-1	4010	600	37-1445	37-927	6-504	122-67	122-75	N/R	N/R	125-65	.035
R84035-2	4010	600	37-1445	37-927	6-504	122-63	122-75	N/R	N/R	125-65	.035
R84037	4011	650	703-59	N/A	6-504	122-63	122-69	N/R	N/R	125-85(30)	.026
R84038	4010	600	37-1445	N/A	6-504	122-67	122-77	N/R	N/R	125-65	.035
R84039	4010	750	37-1445	N/A	6-504	122-79	122-79	N/R	N/R	125-65(15)	.035
R84040	4011	650	37-1447	N/A	6-504	122-60	122-64	N/R	N/R	125-65(15)	.026
R84040-1	4011	650	37-1447	N/A	6-504	122-60	122-64	N/R	N/R	125-65(15)	.026
R84041	4011	800	37-1447	N/A	6-504	122-64	122-90	N/R	N/R	125-65(15)	.026
R84041-1	4011	800	37-1447	N/A	6-504	122-64	122-90	N/R	N/R	125-65(15)	.026
R84042	4011	650	37-1447	N/A	6-504	122-64	122-68	N/R	N/R	125-65(15)	.026
R84044	4010	750	37-1445	N/A	6-504	122-71	122-76	N/R	N/R	125-105(15)	.035
R84044-1	4010	750	37-1445	N/A	6-504	122-73	122-76	N/R	N/R	125-105(15)	.035
R84046	4010	600	703-53	N/A	6-504	122-69	122-76	N/R	N/R	125-65	.035
R84046-1	4010	600	703-53	N/A	6-504	122-69	122-76	N/R	N/R	125-85	.035
R84047	4010	750	37-1445	37-927	6-504	122-75	122-75	N/R	N/R	125-65(15)	.035
R84047-1	4010	750	37-1445	37-927	6-504	122-73	122-75	N/R	N/R	125-65(15)	.031
R84050	4011	650	703-60	N/A	6-504	122-58	122-84	N/R	N/R	125-65(12)	.026
R84412	2300	500	37-474	37-901	6-504	122-73	N/R	134-137	N/R	125-50	.028
R84776	4150	600	37-485	37-910	6-504	122-66	122-73	134-145	134-143	125-65	.028
R84777	4150	650	37-485	37-910	6-504	122-67	122-73	134-150	134-147	125-65	.028
R84778	4150	700	37-485	37-910	6-504	122-69	122-78	134-225	134-226	125-65	.028
R84779	4150	750	37-485	37-910	6-504	122-70	122-73	134-227	134-228	125-65	.028
R84780	4150	800	37-485	37-910	6-504	122-71	122-85	134-229	134-230	125-65	.031
R84781	4150	850	37-485	37-916	6-504	122-80	122-78	134-231	134-161	125-65	.031
R87448	2300	350	37-749	37-901	6-504	122-61	N/A	134-203	N/R	125-85	.031
R89834	4160	600	37-720	37-912	6-506	122-68	134-39	134-241	134-39	125-65	.031

NUMERICAL LISTING AND COMPONENT PARTS

SECONDARY NOZZLE SIZE OR SPRING COLOR	PRIMARY BOWL GASKET†	PRIMARY METERING BLOCK GASKET†	SECONDARY BOWL GASKET†	SECONDARY METERING BLOCK GASKET†	SECONDARY METERING PLATE GASKET†	PRIMARY FUEL BOWL	SECONDARY FUEL BOWL	THROTTLE BODY & SHAFT ASSEMBLY	VENTURI DIAMETER PRIMARY	VENTURI DIAMETER SECONDARY	THROTTLE BORE DIAMETER PRIMARY	THROTTLE BORE DIAMETER SECONDARY
Pink	(3)	(3)	(3)	(3)	(3)	N/R	N/R	N/R	1-1/4	1-1/4	1-11/16	1-11/16
Pink	(3)	(3)	(3)	(3)	(3)	N/R	N/R	N/R	1-1/4	1-1/4	1-11/16	1-11/16
Black	(3)	(3)	(3)	(3)	(3)	N/R	N/R	N/R	1-1/2	1-1/2	1-11/16	1-11/16
Purple	(3)	(3)	(3)	(3)	(3)	N/R	N/R	N/R	1-1/2	1-1/2	1-11/16	1-11/16
.026	(4)	(4)	(4)	(4)	(4)	N/R	N/R	N/R	1-5/32	1-3/8	1-3/8	2
N/R	108-33	108-29	N/R	N/R	N/R	134-103	N/R	112-2	1-3/8	N/R	1-11/16	N/R
.032	108-33	108-29	108-33	108-29	N/R	134-103	134-104	112-16	1-1/4	1-5/16	1-9/16	1-9/16
.028	108-33	108-29	108-33	108-29	N/R	134-103	134-104	112-17	1-1/4	1-5/16	1-11/16	1-11/16
.031	108-33	108-29	108-33	108-29	N/R	134-103	134-104	112-22	1-5/16	1-3/8	1-11/16	1-11/16
.031	108-33	108-29	108-33	108-29	N/R	134-103	134-104	112-18	1-3/8	1-3/8	1-11/16	1-11/16
.031	108-33	108-29	108-33	108-29	N/R	134-103	134-104	112-21	1-3/8	1-7/16	1-11/16	1-11/16
.031	108-33	108-29	108-33	108-29	N/R	134-103	134-104	112-19	1-9/16	1-9/16	1-3/4	1-3/4
N/R	108-33	108-29	N/R	N/R	N/R	134-103	N/R	112-14	1-3/16	N/R	1-1/2	N/R
Black	108-33	108-31	108-30	108-30	108-27	134-122	134-105	112-92	1-1/4	1-5/16	1-9/16	1-9/16

Footnotes

(1) 122-80 Choke Side; 122-90 Throttle Side
(2) Model 2010 Airhorn Gasket is Available Under Part Number 108-75
(3) Model 4010 Airhorn Gasket is Available Under Part Number 108-63
(4) Model 4011 Airhorn Gasket is Available Under Part Number 108-64
(5) Main Body Gasket
(12) 125-85 Secondary
(13) 125-105 Primary
(14) 125-85 Primary
(15) 125-65 Secondary
(16) 6-511 Primary
(17) 6-510 Secondary
(21) 125-65 Primary
(22) 125-35 Secondary
(24) 25R-475A-13 Early versions must use 108-29 to seal pump passage.
(29) 122-75 Diaphragm side; 122-80 Throttle Lever side
(30) 125-25 Secondary
N/A Not Available
N/S Not Serviced
N/R Not Required
†NOTE: Gasket Part Numbers now have a (-2) suffix to denote 2 gaskets per package. For example: 108-38-2.

HOLLEY REFERENCE CHARTS

HOLLEY STANDARD MAIN JETS

Jet No.	Drill Size	Jet No.	Drill Size
40	.040	71	.076
41	.041	72	.079
42	.042	73	.079
43	.043	74	.081
44	.044	75	.082
45	.045	76	.084
47	.047	77	.086
48	.048	78	.089
49	.048	79	.091
50	.049	80	.093
51	.050	81	.093
52	.052	82	.093
53	.052	83	.094
54	.053	84	.099
55	.054	85	.100
56	.055	86	.101
57	.056	87	.103
58	.057	88	.104
59	.058	89	.104
60	.060	90	.104
61	.060	91	.105
62	.061	92	.105
63	.062	93	.105
64	.064	94	.108
65	.065	95	.118
66	.066	96	.118
67	.068	97	.125
68	.069	98	.125
69	.070	99	.125
70	.073	100	.128

HOLLEY POWER VALVES

SINGLE STAGE POWER VALVES (includes gasket)

Standard Flow

Part Number	Opening Vacuum (in Inches of Hg.)
125-25	2.5
125-35	3.5
125-45	4.5
125-50	5.0
125-65	6.5
125-75	7.5
125-85	8.5
125-95	9.5
125-105	10.5

High Flow

Part Number	Opening Vacuum
125-125	2.5
125-135	3.5
125-145	4.5
125-155	5.5
125-165	6.5
125-185	8.5
125-1005	10.5

TWO-STAGE POWER VALVES (includes gasket)

Part Number	1st Stage Opening (in Inches of Hg.)	2nd Stage Opening (in Inches of Hg.)
MODEL 4160		
125-206	12.5	5.5
125-207	10.5	5.0
125-208	10.5	5.5
125-213	11.5	5.0
MODEL 4175		
125-209	11	6.0
125-210	9	2.5
125-211	10.5	5.5
125-212	12	6.5
125-215	10	6.0
125-216	8	1.5
125-217	10	4.0
125-218	11	5.5
MODEL 4360		
125-200	9	5.0
125-201	8	5.0
125-202	8	4.0
125-203	8.5	5.5
125-204	8	3.0
125-205	9	3.0

HOLLEY ACCELERATOR PUMP DISCHARGE NOZZLES

TUBE TYPE

Part No.	Hole Size
121-25	.025
121-28	.028
121-31	.031
121-35	.035
121-37	.037
121-40	.040
121-42	.042
121-45	.045

STRAIGHT TYPE

Part No.	Hole Size
121-125	.025
121-128	.028
121-131	.031
121-132	.032
121-135	.035
121-137	.037
121-140	.040
121-142	.042
121-145	.045
121-147	.047
121-150	.050
121-152	.052

ANTI-PULLOVER DISCHARGE NOZZLE

Part No.	Hole Size
121-225	.025
121-228	.028
121-231	.031
121-237	.037
121-240	.040
121-242	.042
121-245	.045
121-247	.047
121-250	.050
121-252	.052

CLOSE LIMIT MAIN JETS

Jet #	Jet #	Jet #	Jet #	Jet #
352	432	512	592	672
362	442	522	602	682
372	452	532	612	692
382	462	542	622	702
392	472	552	632	712
402	482	562	642	722
412	492	572	652	732
422	502	582	662	742

SECONDARY DIAPHRAGM SPRING KIT PN. 20-13

COLOR	RELATIVE LOAD
White	Lightest
Yellow	Lighter
Yellow	Light
Purple	Medium Light
Plain	Medium
Brown	Medium Heavy
Black	Heavy

If the engine bogs under hard acceleration (too much carburetion at slow engine speed), installing a heavier spring in the secondary diaphragm housing may provide a cure. Install springs one step heavier at a time. Using lighter springs will open the secondaries earlier and may help acceleration in cars that are sluggish up to 2500 to 3000 RPM (when caused by insufficient carburetion at low engine speed).

HOLLEY SECONDARY METERING PLATES

Main Hole	Idle Hole	Order By Part No.	Part Stamped
.052	.026	134-7	7
.052	.029	134-34	34
.055	.026	134-3	3
.059	.026	134-4	4
.059	.029	134-32	32
.059	.035	134-40	40
.063	.026	134-5	5
.064	.028	134-18	15
.064	.029	134-30	30
.064	.031	134-13	13
.064	.043	134-33	33
.067	.026	134-8	8
.067	.028	134-23	23
.067	.029	134-16	16
.067	.031	134-9	9
.067	.035	134-36	36
.070	.026	134-6	6
.070	.028	134-19	19
.070	.031	134-20	20
.070	.033	134-41	41
.071	.029	134-35	35
.073	.029	134-39	39
.073	.031	134-37	37
.073	.040	134-17	17
.076	.026	134-10	10
.076	.028	134-22	22
.076	.029	134-43	43
.076	.031	134-12	12
.076	.035	134-53	3
.076	.040	134-28	28
.078	.029	134-38	38
.078	.040	134-52	52
.079	.031	134-11	11
.079	.035	134-24	24
.081	.029	134-44	44
.081	.033	134-49	49
.081	.040	134-21	21
.081	.052	134-31	31
.081	.063	134-39	29
.082	.031	134-46	46
.086	.043	134-25	25
.089	.031	134-47	47
.089	.037	134-55	5
.089	.040	134-27	27
.089	.043	134-26	26
.093	.040	134-54	4
.094	.070	134-15	15
.096	.031	134-50	50
.096	.040	134-45	45
.098	.070	134-14	14
.113	.026	134-42	42

HOLLEY NEEDLES & SEATS

VITON INLET

Seat Size	Order By Part No.	Type
.097"	6-506	Adjustable
.097"	6-507	Adjustable
.097"	6-508	Adjustable
.097"	6-517	Adjustable
.110"	6-504	Adjustable
.120"	6-518	Adjustable
.101"	6-520	Adjustable
2mm	6-512	Model 5200
.0785"	6-511	Non-Adjustable
.100"	6-516	Non-Adjustable
.110"	6-510	Non-Adjustable
.110"	6-514	Model 4360
.97"	6-513	Off-Road

STEEL INLET

Seat Size	Order By Part No.	Type
.97"	6-501	Adjustable
.110"	6-500	Adjustable
.120"	6-502	Adjustable
.130"	6-515	Adjustable
.150"	6-519	Adjustable